Just the Facts

Investigative Report Writing

THIRD EDITION

Just the Facts

Investigative Report Writing

MICHAEL BIGGS
Long Beach City College

PEARSON

Prentice
Hall

Upper Saddle River, New Jersey 07458

CIP can be obtained from the Library of Congress.

Editor-in-Chief: Vernon R. Anthony
Senior Editor: Tim Peyton
Editorial Assistant: Jillian Allison
Marketing Manager: Adam Kloza
Managing Editor: Mary Carnis
Manufacturing Buyer: Cathleen Petersen
Production Liaison: Ann Pulido
Production Editor: Rebecca Anderson/Pine Tree Composition, Inc.
Composition: Laserwords
Senior Design Coordinator: Miguel Ortiz
Cover Designer: Miguel Ortiz
Cover Image: Rob Koeberer, Getty/Aurora
Printing/Binding: Bind-Rite Graphics
Cover Printer: Phoenix Color Corp.

Pearson Prentice Hall™ is a trademark of Pearson Education, Inc.
Pearson® is a registered trademark of Pearson plc
Prentice Hall® is a registered trademark of Pearson Education, Inc.

Pearson Education LTD.
Pearson Education Australia PTY, Limited
Pearson Education Singapore, Pte. Ltd.
Pearson Education North Asia Ltd.
Pearson Education Canada, Ltd.
Pearson Educación de Mexico, S.A. de C.V.
Pearson Education—Japan
Pearson Education Malaysia, Pte. Ltd.
Pearson Education, Upper Saddle River, New Jersey

10 9 8 7 6 5 4 3 2 1

ISBN-13: 978-0-13-134763-2
ISBN-10:   0-13-134763-2

To the KCBs and Tucker

Contents

CHAPTER 10 Issues in Writing 145

Preface

I am glad I have friends and colleagues who care as much about helping investigators develop and improve their report writing skills as I do. Shortly after the first edition of this book was published in 2001, I started receiving feedback about the text—not only kind words and support but also a lot of ideas to make it better. These suggestions not only covered a wide variety of areas but also supported the basic premise of the first edition—that there is a need for an entry-level report writing guide that combines some of the basics of investigation with how to write about them.

The third edition of *Just the Facts* brings together the good points of the first and second editions with several new pieces of information as the result of the feedback of many people. More than two decades of practice with the rules of writing and exercises—with continual feedback from students, police officers who have attended the class and put these techniques into practice, and other report writing instructors—have convinced me that this system works.

This workbook is an attempt to meet the needs of report writing students by establishing fundamental guidelines for investigative reports through a set of rules that are easy to understand and apply in any situation. By following these rules, a student can break down each major component of investigative report writing into its simplest form and examine it for weaknesses. These weak points can then be corrected, resulting in immediate improvements.

Each chapter is devoted to a major component of the report writing process and builds on the previous learnings. Each chapter provides an overview of what will be discussed and then offers a list of key terms that will be covered in it. Each chapter concludes with a summary, followed by a short review, a set of exercises to build on the material that the chapter covered, and a ten-question quiz. The exercises are designed to build confidence and reinforce the topics just covered. This edition includes two new chapters, an appendix with three report condensing exercises, an appendix with blank report forms, and a glossary.

Chapter 1 describes the philosophy of police report writing. It provides a historical perspective of who the first report writers may have been, how police reports have evolved into being the source of all data used in the cops-and-robbers business. It discusses records management systems and the uniform crime reporting program. It also discusses the mechanics of writing, report writing technology, the dictation process, and the report writing approval process. The chapter concludes with a discussion of the need for effective report writing and the reason this book was written.

Chapter 2 covers the basics of investigation, the goal of an investigation, and what steps to take in initiating an investigation. The chapter also lists the qualities that superior investigators demonstrate and provides the reader with the opportunity to compare his or her traits to those qualities.

Chapter 3 is all about note taking, field notebooks, and the desired outcome of note taking opportunities. The discussion of notebook types and which one is right for each investigator should prove helpful to those new to the business.

Chapter 4 provides the framework for writing good investigative narratives and how to overcome spelling problems. This chapter will give any investigator the skills to write effective reports.

Chapter 5 clearly defines the players in a report, including the reporting party, victim, suspect, and witnesses. The chapter not only identifies the people in a report, but also provides some basic guidelines for describing suspects and property.

Chapter 6 covers the purpose and uses of crime reports and how to effectively complete the face sheet and a complete narrative.

Chapter 7 addresses when an arrest report is needed and how to complete one. It offers suggestions for completing a report when no set format is available.

Chapter 8 deals with the relationship between the interview process and the role it plays in report writing. The chapter looks at verbal and nonverbal clues for the interviewer, describes how these clues can be used to establish a behavior baseline, discusses the need for preparation and how and when to conduct the interview, reviews the Miranda admonishment, and explains how to write the results of the interview.

Chapter 9 provides the fundamentals of writing search warrants and begins by explaining the parts of a search warrant and the process to be followed in obtaining one.

Chapter 10 acts as a summary and brings to the forefront several report writing problems and ways to solve them.

Acknowledgments

No workbook like this comes from a single source, and I want to thank those who gave their time reviewing the manuscript and providing feedback on the text. This includes Karin Dudash, Cameron University, Lawton, OK; Carolyn D'Argenio, Mohawk Valley Community College, Rome, NY; Arnold Waggoner, Rose State College, Midwest City, OK; and William Van Cleve, California State University—Fullerton, Fullerton, CA. I also want to thank Police Chief Tim Grimmond, El Segundo Police Department, retired, and Captain John Rees, La Habra Police Department, retired, for their assistance in providing the report forms used as examples in the text. Last but not least, I would like to acknowledge and thank the hundreds of others whose comments during classes and breaks helped me focus on some of the finer points of writing. I hope this helps.

Mike Biggs

1

Why We Write Police Reports

KEY POINTS

Many times it is easier to understand how to do something if you know and appreciate why you are doing it. This chapter is designed to provide some background on why we write police reports. It covers such things as the mechanics of writing, records management, and the role reports play in crime statistic reporting and why effective report writing is so important in law enforcement. As well, the chapter includes a brief discussion on why this book was developed and how it can be applied to investigative report writing training.

KEY TERMS

Automated reporting systems
Body of data
Bubble in system
Centralized approval
Check the box
Computer
Dictating
Fill-in-the-blank
Fixed terminal
Forced-choice features
Grammar check
Narrative part
National Incident Based
  Reporting System

New York State Incident-Based
  Reporting program
Optical character recognition
Part I crimes
Pull-down menus
Records management system
Spell check
Template
Touch-screen features
Transcribes
Uniform Crime Reporting
  Program
Voice dictation

BACKGROUND

Who was the first police officer? When was the first investigative report written? Are the cave drawings and inscriptions the Mayan and Inca historians left us an effort to tell us about a crime trend? Do the picture script hieroglyphics of the ancient Egyptians tell the story of a crime against the pharaoh? We may never know, but what we do know is that written police reports have been around for a long time. Police reports are the way we document the history of law enforcement and the activities in which its members are involved.

If the law enforcement community is anything, it is a great record keeping institution. In a given year, most agencies can specifically identify how many traffic tickets were written, how many arrests were made, and even how many minutes a specific officer spent doing various activities. This kind of data is part of the lifeblood of law enforcement planning and staffing activities. It is part of the data that law enforcement planners use to project beat configurations, hours of work, and how many officers should be deployed. There is, however, another aspect of the record keeping and data storage process, and that is what to do with all of the statistics and information. Should it be shared with the public and other law enforcement agencies? Does it have any value in trying to solve crimes? If the information is looked at on a grand scale, does its potential value increase?

RECORDS MANAGEMENT SYSTEMS

It may be helpful to understand how the information in police reports, especially investigative reports, is generated, collected, and ultimately used. The crime and statistical information that makes up the **body of data** has its beginnings at a local level. By this I mean that from a municipal or county agency perspective, statistics and information flow forth from an area, beat, or reporting district. At the state or federal level, these initial reporting zones might be a city, county, or field office. Nonetheless, the information soon makes its way to a reportable point, which is usually some form of **records management system**. These systems can take many forms, from a veteran officer's memory, to a file cabinet, to a large Rolodex, or to complex automated databases. The common thread in any of these systems is that someone needs to be able to extract information about victims, suspects, what happened, where it happened, and when it happened and then provide it in a usable format to those with a need and right to have it. Within the past couple of years, I have had two instances when my request for crime statistics was answered with, "We don't keep anything official, but I can tell you what happened." Was the information provided to me accurate? It is difficult to tell with certainty, but it appears that the information I received was correct.

Information is put into various records management systems in a number of ways. Clerks manually enter some of the information; at other times, the investigators who generate it input it. Other systems are linked to automated reporting systems whose database automatically includes the information as the report writer enters it onto the report form. The information collected and reported has traditionally been in a quantitative format and is typically shown as pure numbers or as a percentage change from years past. This kind of statistical reporting by city, county, state, and, in some cases, regions, was seen by law enforcement professionals in years past to be very helpful. In 1929, the International Association of Chiefs of Police (IACP) developed the **Uniform Crime Reporting Program** as a way to standardize the information. The leaders of IACP recognized the value in using such information to track and monitor crime trends, evaluate training programs, and assess the effectiveness of laws and policies as they were implemented. Out of this initiative, the Federal Bureau of Investigation, with authorization from Congress, effectively became the owner and operator of the Uniform Crime Reporting Program and receives crime statistical information from participating law enforcement agencies.

The Uniform Crime Reporting Program collects and tracks statistics on what is known as the eight **Part I crimes**, which include Murder, Rape, Robbery, Aggravated Assault, Burglary, Larceny-Theft, Motor Vehicle Theft, and Arson. Every month, nearly 17,000 law enforcement agencies report their statistical information to the FBI on a voluntary basis, where it is added to the historical files. The information then becomes the basis for the yearly report, known as the Summary Data Reporting System. In 1982, the FBI and the Bureau of Justice Statistics, which is part of the U.S. Department of Justice, began a five-year review of the Uniform Crime Reporting Program that resulted in a substantial redesign culminating in the creation of the **National Incident Based Reporting System** (NIBRS). The NIBRS not only captures the summary data it always has but also includes detailed information on an expanded set of crimes.

While the old Uniform Crime Reporting Program captured limited information about a crime event, the NIBRS is designed to obtain more specific details about each reported crime including the date, time, location, and circumstances of the incident; the sex, race, and age of the victim and the offender; any relationship information about the suspect and victim; whether drugs or weapons were involved; and evidence of whether the crime was motivated by bias.

While full participation in and use of the NIBRS will help law enforcement track, analyze, and combat crime, the problem is that not all uniform crime reporting agencies are supplying data at the level NIBRS needs to be effective. In fact today, a multitude of agencies still use different report forms, writing formats, and approval processes. While these agencies have undoubtedly conducted quality investigations and, in many cases, achieved good results, from a national perspective, the entire reporting system is not as smooth and well developed as it could be. One example of law enforcement moving in the right direction in this area is the **New York State Incident Based Reporting Program** (NYSIBR). The New York model sets a good standard that includes:

- Using a standardized incident report.
- Using a standardized arrest report.
- Viewing the incident as a set of related components.
- Looking at the relationships between suspects, victims, and property.
- Studying victim and suspect demographics.
- Identifying the use of weapons as well as drug and alcohol involvement.

THE MECHANICS OF WRITING

As good as these programs are, they are only as good as the information put into them, and that leads the discussion to how police reports are prepared. At the risk of being too simplistic, I suggest there are two basic parts to any police report: The fill-in-the-blanks part and the narrative part. The **fill-in-the-blank** part includes the prompted, generic information such as name, address, and phone number. The fill-in-the-blank information can also take the form of **check the boxes**, forced choice, and fill-in-the-bubble options—generally anything that is a checklist. The value of this kind of information is that it becomes a consistent stream of data for the agency's case management system. This kind of check the box, forced choice, and **bubble in system** may also have value in entering data in the case management system through the use of optical character recognition.

The **narrative part** occurs when the investigator tells the story of what happened, what actions were taken to solve the crime, and what—if any—evidence was collected. In short, this is the meat and potatoes of the investigative work, and learning how to do it well remains the most difficult part of the report writing process.

REPORT WRITING TECHNOLOGY

Technology has improved many aspects of law enforcement including report writing. While the goal of implementing technology can be a higher level of officer safety, an improved level of service, time savings, or cost reductions, using automation in report writing can accomplish parts of all of these. The two primary types of automation and technology to impact the report writing function are computers and, to a lesser degree, dictation of reports using tape recorders.

Computers

There is no doubt that computers have had an enormous impact on the cops-and-robbers business. From property and evidence management, to evidence analysis, to statistical data management, computers have been integrated into almost every function of law enforcement. Whether the report writing function has been adapted to the use of computers or vice versa is debatable. What is without question is that there have been a variety of successful cross-applications.

What Is Out There

Probably the most used technique is the computer often placed in a patrol car. This too has seen some variety in that some agencies use a portable laptop and others use a fixed terminal that is hardwired into the car.

A common approach to writing reports in the field is to have field officers document their preliminary investigations as soon as possible by using one of the many and varied software systems in use. There is a wide variation in the types of software used because some agencies have developed their own and others have customized off-the-shelf versions. Each system is usually tied to the specific needs of the agency's records management system, and so the chance that a single common approach will be developed is not very likely. A records management system allows departments to electronically enter data into their agency files, search for specific items, update necessary cases, and do many other tasks electronically. It also allows for the reduction of paper files because the majority of information is stored electronically. Currently used software allows officers to complete initial and supplemental reports as well as traffic investigation reports.

Special Features

The features that these various systems use include template report forms, the use of forced-choice data fields, pull-down menus, grammar check, spell check, and other touch-screen processes.

Templates

With a **template** feature, writers can start by calling a blank form to the screen with a click of the mouse. This form can be interfaced with various city or county databases to instantly verify, correct, and add phone numbers, zip codes, reporting districts, beat or section numbers, and other data that are used in the statistical management of the agency. Using templates also allows an officer to work on several reports over a period of time and ensure that they are neat and readable upon completion. Templates also ensure that all officers are using the same form and format for their reports.

Forced Choices

Forced-choice features will allow a writer to check the box or fill in the bubble of the most likely modus operandi (M.O.) factors for certain crimes. This choice can then be downloaded into the records management system and allow for regional crime trend analysis.

Pull-Down Menus

Pull-down menus are much like a forced-choice feature and allow the writer to choose the most appropriate selection from the many that are offered by clicking on it, which enters it in the appropriate field. Use of this feature will provide consistency and correct spelling and will allow for the repetitive information in a report to be quickly identified and entered in the proper locations.

Grammar Check

Grammar check helps writers by checking the proposed language of a narrative against standard usage and offering suggestions when a questionable situation occurs. For agencies that specify first person and active voice, the grammar check can almost guarantee that officers will be prompted to reconsider their word choices when their work falls outside these parameters.

Spell Check

Spell check will help officers overcome one of the biggest challenges they face, misspelling words. Spell check can be customized to add names of department personnel and streets to ensure the accuracy of the report. Spell check is not foolproof, however. The drawback is that if you choose an unusual word that is not in the word-processing program's dictionary you are using, you will be notified that it is being questioned, but you may not be given a correct option to choose.

Touch Screen

Touch-screen features allow for rapid selection of common data. Their use in report writing could be in the selection of the type of report, date and time, distribution, and recommended follow-up.

OPTICAL CHARACTER RECOGNITION

Optical character recognition (OCR) was first patented in 1929 by G. Tauschek in Germany. Typically, OCR translates scanned images of handwritten or typewritten words into computer-editable text. This is very useful and cost effective in entering data into the records management system and is having some success in law enforcement. OCR is fairly well developed as it pertains to typewritten text and reports, but when it comes to printed or cursive handwriting and even typewritten reports that contain a large number of characters, it is a developing technique.

DICTATION

Another technological option that law enforcement has used for years is dictation. The dictation process for report writing has two components, dictating and transcribing. **Dictating** is the part of the system in which someone reads or says something that another person writes, usually with a specific code or a specialized machine, and then later **transcribes** or writes it down. The transcribed information is then formatted and printed. The law enforcement community has used dictation for at least 35 years after adopting it from the business world. While dictating reports has helped streamline the investigative report writing process for some patrol officers and detectives, it has had mixed results for a number of reasons. Two types of dictation are voice dictation and person-to-person dictation.

Voice Dictation

Voice dictation uses a preprogrammed software feature recognition that recognizes the speaker's voice. The report composer merely dictates the report and the software, after identifying and recognizing the voice, converts the report to transcribed text.

Person-to-Person Dictation

While some agencies have used **automated reporting systems** and have found them helpful in solving some investigations, not all departments are able to implement such a system. They have the option to use person-to-person dictation described earlier to explain the dictation process.

Benefits of Dictation

The benefits of a dictation system for law enforcement are many. A prime consideration is that it can save field officers and investigators a lot of time. Most people can talk faster than they can write, and when this is multiplied over many shifts by many officers, the time saved by dictation can be substantial. Another benefit is that the finished product is typewritten and usually neater in appearance than a handwritten report. This has a second benefit for those who use the initial reports because they are easier to read and hopefully make follow-up work easier to do. Problems such as being unable to distinguish numbers and letters in license plates, addresses, and phone numbers should be eliminated with an accurately dictated, transcribed, and typed report. Another benefit of a solid dictation process is that the system can be connected to an automated record keeping system and save a lot of time indexing names and addresses. Many departments enter every name and address that appear in a police report into their records bureau master file. This allows investigators to check names at a later date and see every occasion that an officer has been involved with a particular person or address. This is possible by capturing information in various fields when the report is being transcribed. The information captured in these fields is automatically entered into the records data base as the report is prepared, thereby saving time and giving field officers up-to-date information on a very timely basis.

Problems with Dictation

Just as the benefits of dictation cover a large area, the problems encountered can be tremendous. First, the equipment used must be in good working order. Poorly maintained tape recorders, old tapes, and poor microphone or telephone connections can cause problems. Second, loss of power or weak batteries can lead to an incomplete or inaudible tape. Third, background noise can sometimes distort the reporters voice. Fourth, some investigators forget to provide the framework for the report such as spelling names, providing punctuation, or giving specific directions as to the gramatical structure. Last, but not least, some people talk too loud, too soft, or too fast and thereby record an unuseable tape. While these might seem like minor problems, any one of them can bring the dictation process to a stop, thereby contributing to a less than smooth process.

DEPARTMENT SYNERGIES

Automated reporting systems rarely stand alone in this day and age. They are usually hooked in with one or more other department wide systems such as computer aided-dispatch or automated records management. As these systems are refined and interconnected, the time and money savings can be substantial. When department budgets are constantly scrutinized and police managers are expected to cut costs, an automated report writing system might be part of the answer.

COST-EFFECTIVENESS

When officers become proficient at completing reports in an automated reporting system, the amount of time they will need to write a report will usually be reduced. This will translate to more officers being available for routine calls or directed patrol activities, which in a large enough department can result in fewer officers being needed to staff shifts. A second cost saving can occur when record systems are automatically updated. This efficient and accurate method of transferring data from reports into a database can be an opportunity to reduce staffing in records bureaus. Reproduction and storage costs can also be reduced when information is stored electronically.

A fourth area of cost savings can occur if the public can access the automated report system. Anyone with access to the web could complete basic reports with no suspect information. Citizens could connect to an agency's web site, download a report form template, and after completing it, email it or send it to the department. Again, less time spent by officers on these kinds of reports will translate to more effective use of time.

TRAINING

There is no doubt that some training will be needed to get officers comfortable with this type of technology and these kinds of changes to the traditional way of doing police work. There is also a real possibility that in the initial use period, there will be an increase in the amount of time needed to complete reports. This should decrease with experience and system familiarity.

THE APPROVAL PROCESS

Even with the aid of computers, the task of the supervisor to review and approve reports or send them back for correction is still key to a successful system. In the traditional paper-based report writing systems, officers would complete a handwritten report and meet with a supervisor in the field or turn it in at the end of the shift. The supervisor would then read the report and approve it or return it to the officer for correction. The automated systems provide several options to the field officer to send reports to a supervisor. If the supervisor is on line, access is immediate and constant. The supervisor's computer can even be set up to offer a noise prompt or page to a cell phone when a report is in his or her in basket. Some officers might save the report to a floppy disk and give it to the supervisor in the field or at the station. The supervisor would then bring the report up on a computer to review and electronically approve or print a hard copy out and approve it. Notification to the writer can also be sent electronically.

Another method for getting the report to a supervisor is to take the car laptop into a station or headquarters and use a docking station to download the report into the system. Supervisors then can gain access to the report and carry out their approval function.

Centralized Approval

Some departments have opted to use **centralized approval** for the report review and approval process in specialized units. In this model, a group of officers and supervisors undergoes special training in the area of report review and correction and, in essence, becomes the department experts in the field. All reports are sent to this group, whose responsibility it is to review and ultimately approve all reports written by the department's officers. Many times this group of experts many times are also tasked with working with officers who are having report writing problems and giving them the necessary training needed to improve their skill set.

The benefits of this type of model are many. First, it creates a centralized and well-trained staff to handle this important task. It also provides consistency to the field officers who can expect their work to be viewed objectively. The report review and approval team is also in a position to identify training issues, not only report writing needs for individual officers but also department wide training issues. This model also allows the department experts to make suggested corrections for the originating officer to consider before resubmitting the report for approval. Another benefit this model provides is that it frees up the field supervisors from what is often a laborious and time-consuming task.

THE NEED FOR EFFECTIVE REPORT WRITING

The law enforcement community clearly recognizes report writing as an important job skill. This is shown by the inclusion of report writing training in local, state, and federal basic officer and agent training programs. However, this allocation of training time and dollars has not resulted in a uniform or consistent approach. A check of several law enforcement standards and training agencies in the United States shows as few as 4 hours to as many as 40 hours devoted to the topic in basic officer courses. This is supplemented with on-the-job in-field training, yet some in law enforcement management still consider report writing to be a topic deserving more attention.

Some years ago, one law enforcement standards and training agency surveyed the top law enforcement managers in the state and asked them to identify the top three problems facing their agencies. The results were hardly surprising; they said too many officers were being killed in the line of duty, the general knowledge of and application of the law was less than desired, and despite a comprehensive training program, many officers had report writing problems.

Why does report writing remain a problem for law enforcement? I think that a number of factors contribute to the cause, which in no particular order include:

1. **Not all training programs get the job done.** An outstanding lesson plan delivered poorly may not prepare the student to actually do what he or she was taught to do.
2. **Not all students are created equal.** I think it is the instructor's job to find the best way to get through to a student, but with that said, not all students achieve superiority.

3. **Some field training programs are still dealing with the issue of what I call the negative reinforcement of training.** This occurs when a new officer meets the field-training officer for the first time and is told, "forget what you learned in the academy, I'm going to show you how it is really done."

4. **Report writing genealogy.** This happens over several generations of incoming officers and can have both a positive and negative aspect. When an officer develops a reputation as a good report writer, others look to her or him as a mentor. The struggling officers tend to emulate the writing style of the "good writer" and begin to use the same phraseology and word choice. If they become good writers, the cycle repeats itself. Therefore, someone four or five generations down the line could write just like the first good writer. How "good writers" are defined is a function of the individual agencies.

5. **Lack of consistency among supervisors.** Very few report writing training programs include a component for the first line supervisors and management team. The focus is usually on the "problem" officer or agent and some sort of remedial training program. I think this is an approach that will allow the problem to continue. Supervisory review needs to be thorough and in sync with others.

6. **Law enforcement free agency.** Back in the day, officers tended to serve their entire career with a single department. Today, it is not uncommon for experienced officers and agents to move around to get better pay and benefits. With this phenomenon comes new ideas and techniques that are carried from department to department, and this can include report writing techniques—both good and bad.

7. **Trial by fire.** Plea bargaining is a fixture in our criminal justice system. This affects the report writing process because it removes a critical part of the review process—a challenge by defense attorneys. When officers must back up their written work by facing someone in an adversarial role whose job is to represent the accused, lessons learned may be hard, but they are lasting. When cases are disposed of without an officer testifying, a valuable part of the learning process does not take place. This makes the supervisory role in reviewing and correcting young officers' reports even that much more critical.

Even with advances in technology that make the report writing function easier to achieve, the problem of teaching narrative writing skills to investigators remains. The narrative section of the report is really where the story is told. Things are explained, the details of not only what but how it happened are shown. This can be the most challenging and most rewarding part of being an investigator because long after your effort is complete, the written record of your work lives on.

WHY THIS BOOK

I think it is safe to say that there is a need for a solid, easy-to-use methodology to teach police report writing. I also believe that once a basic set of writing skills is mastered, it becomes easier to document a law enforcement event. What changes for the young officer or investigator is not the writing but the parts of the investigation. The investigating is the part of the process that presents the learning curve challenges. Consider this saying that is food for thought: "Good writers can write about anything." So the challenge now becomes how to teach people to become good police report writers, and many criminal justice instructors and professionals have an opinion about how to do it. Some of the ideas that I have heard over the years have merit, but I believe they complicate the process of teaching a student to become a successful report writer. They include:

1. **You have to teach basic grammar.** I disagree. A prospective police officer or investigator must bring a certain skill to the game. In this skill set must be the ability to write. The student needs to know verb tenses, subject-verb agreement, how to put a sentence together, and how to structure a paragraph. It is not the purpose of this book to teach grammar, nor do I believe it is the job of a police report writing instructor to teach someone how to write. I believe it is the job of such instructors to teach people who know how to write, how to write a police report. This book supplies a tried and proven system to do just that.

2. **Teach spelling and make sure students use a dictionary.** I absolutely agree that police officers should spell words correctly in their reports, and I absolutely disagree that a dictionary is the way to achieve this. Dictionaries show us what a word means, how it should be used, and in some cases alternate words that mean the same thing. The problem with using a dictionary to find the correct spelling of a word is that in many cases, you need to know how to spell it to find it in the dictionary. The other issue with this concept is that not everyone has trouble spelling the same words; we all have our own personal spelling demons. The good news on this one is that it is easily corrected and all that is needed is a personal "cheat sheet" of the words an officer has trouble with. Just write down the words you are having trouble with on a card and keep it in your notebook. Add to it as needed. This takes care of the problem.

3. **Use automated teaching materials such as web sites that have writing scenarios.** While I fully appreciate the use of innovation in teaching, I feel the basics of police report writing are still best introduced and reinforced the old-fashioned way—with pencil and paper. Identifying and publishing a list of web sites that are unavailable a short

time later would be detrimental. The use of video scenarios, in-class actors, and staged situations will give students the chance to use multiple skills in the development of their investigation and writing ability. I feel that each instructor should have the freedom to use special scenarios involving live people to give that one-on-one contact students need to practice what they are learning.

My philosophy is to teach a system that can be applied in multiple situations as well as identify and correct report writing deficiencies. I also think instructors need to tie report writing to a second needed skill, how to investigate, and using the synergies of these two independent but interrelated skills creates a learning environment that is challenging and practical. I fully realize that when it comes to law enforcement training procedures and techniques, there is usually more than one right way to do something. This is absolutely true when it comes to teaching report writing and certainly begs the question of what to teach first—note taking or describing property—and then in what order should the other things that need to be taught follow. I would be hard pressed to say that the progression of information in this text is "the best," but I will say that it makes sense to me, it builds on previously taught skills, and at the end of the semester, the student has been exposed to all of the material. I do not look at the semester on a week-by-week basis but as a whole entity. I have heard from some instructors who move from chapter to chapter in a somewhat random order, and they report good results. I have also heard from instructors who track their lesson plans in the same order of the chapters, and they also report good results. The bottom line on this one for me is not what works best for some instructors but what works best for the student because that is why we teach.

I also like to evaluate a student's work product using the "Watch Commander's Test." In any police department there is an established routine that is based on the workload, the people involved, and the legal timelines. These things combine to be known in general terms as "getting the work out," and one of the duties of the watch commander is to review and approve reports. Some watch commanders approve only perfect reports and return all others to the writer, even if the correction needed is very minor. Other watch commanders are willing to make small corrections that do not change the facts or meaning of the report. This standard is what I refer to as the "watch commander's test." If I were the watch commander and a student gave me his or her report to review, the application of the test is simple. If I would approve it, the report is good enough, and if I would not approve it, the report needs more work. I agree that this is a subjective way of looking at a student's work, but it is also a real-world test of his or her ability. After all, isn't that the point of this type of training?

SUMMARY

When the first police report was written is open to debate, but we do know that many pieces of the criminal justice system connect to the report writing process and that automation and technology have influenced this critical piece of an investigator's daily work. Police reports form the basis of crime statistics and are used by nearly 17,000 agencies to provide the information contained in the uniform crime reports. Law enforcement agencies are continually looking for ways to improve the quality and mechanics of report writing and have welcomed technology to the fold to further this effort. Word-processing programs, template-driven systems, touch-screen systems, and optical character recognition are several of the newer items to be considered and tested. Another piece of the report writing process is getting the information in the report into a database, and this is where automated reporting systems, computer-aided dispatch systems, and records management systems come into play. The success of these systems has a direct correlation to the report review and approval process and depends on the report approvers' work to be timely and accurate. How new officers and investigators learn to write can be as varied as the imaginations of the trainers charged with this responsibility. Simply put, there is no single right way to teach someone to write a police report. What is important is that whatever method a new field officer learns, it needs to be simple enough to cover the day-in, day-out basic reports the officer must complete yet complex enough that it can be adapted to any crime or situation that requires written documentation.

REVIEW

1. The International Association of Chiefs of Police established the Uniform Crime Reporting System in 1929.
2. The FBI is charged with managing the collection of crime statistics for the eight Part I crimes.
3. The National Incident Based Reporting System is an attempt to expand the value of the uniform crime reports by including detailed information on an expanded set of crimes.
4. It takes two or more people to have an effective dictation process.
5. Automated reporting systems almost always depend on supporting systems such as computer aided dispatch and automated records management.
6. A solid report review and correction process is key to a successful report writing program.
7. Knowing how to write is key to learning how to write a police report.
8. A customized spelling cheat sheet is a better option than a dictionary.

EXERCISES

1. Visit a library and research to find a criminal case from the 1950s–1960s that received publicity. Have any portions of the police reports been published? What is your opinion of the style of writing, information contained, and completeness?

2. Conduct an Internet search for examples of uniform crime reports and compare these for three or more years. Is the information easy to identify? Do you think the information is useful?

3. Using a software writing program, compose a short report about something you recently did. While writing, deliberately misspell some common words and use poor grammar. Did the software program identify your mistakes and suggest corrections? Do you think this would be helpful if you were preparing an official report?

4. With one or more members of the class, brainstorm ideas that you think would be helpful in a software report writing program. Keep this list in your class notes and review it again at the end of the semester. Are there any changes or additions to make. How many of your original ideas would still be beneficial?

QUIZ

1. What is the name of the organization that developed the Uniform Crime Reporting Program?

2. Which agency is charged with the responsibility to act as the owner and operator of the Uniform Crime Reporting Program?

3. The Uniform Crime Reporting Program collects data on the eight Part I crimes. Name them.

4. Why was the Uniform Crime Reporting Program revised in 1982?

5. What is the revised program called?

6. What are the two main parts of a police report?

7. How does a forced-choice feature make automated report writing easier?

8. What is an effective way to overcome a spelling problem?

9. How does law enforcement "free agency" affect report writing?

10. With regard to the report writing approval process, what is centralized approval?

2

Investigation Basics

KEY POINTS

Investigating crimes and report writing go hand in hand, and a solid understanding of what an investigation is and how one is conducted provides a good foundation to learn about report writing. This chapter provides this foundation by looking at who investigators are, what an investigation is and is not, the goal of an investigation, and how to conduct one. The chapter concludes with a look at the qualities of a good investigator and provides a yardstick to measure your skill set against these qualities.

KEY TERMS

Administrative investigation

Average investigator

Credibility

Criminal investigation

Investigation

Investigators

Superior investigator

Investigative reports do not just happen. They are the formal product of some event that has been brought to the attention of an investigative body or agency and, after a thorough examination, has been documented. Since a good portion of an investigator's time is often spent investigating these events before a report is written, it is logical that any instrument that teaches report writing should also include some discussion about investigators and their investigations.

WHO IS AN INVESTIGATOR?

Investigators are people who look into events or situations to find the facts about what happened. They ask questions and interview people, look at crime scenes, examine documents, collect evidence, develop informants, find stolen or missing property, and develop an understanding of what occurred after reviewing all the available information. Investigators are police officers, deputy sheriffs, security guards, firefighters, claims adjusters, private investigators, personnel specialists, and many other categories of people who are required to possess certain investigative skills and knowledge in order to perform their duties. There is no single set of life experiences or level of education that qualifies someone to be an investigator, but as we will see, there does seem to be some common ground for those who are successful.

WHAT IS AN INVESTIGATION?

An investigation means different things to different people; however, for the purpose of investigative report writing, an **investigation** may be defined as a lawful search for things or people. In each case, the goal is to find the truth. This definition applies to both criminal and administrative investigations.

WHEN DOES AN INVESTIGATION OCCUR?

The media and the war stories of hundreds of investigators have created a commonly held belief that law enforcement representatives are able to investigate anyone at any time for any reason. Few things could be further from the truth because a **criminal investigation** cannot begin until one of three things is present:

1. A crime must have occurred.
2. There must be a reasonable certainty that a crime has occurred.
3. The investigator must be reasonably sure a crime is going to occur.

This means that the investigation and report writing process could occur in one of two ways. The first would begin with the occurrence of a crime, followed by an investigation, and ending with the solution being discovered and proved. In the other sequence, the investigation begins, then the crime occurs, and ultimately the solution is discovered and proved.

After the Crime Has Occurred

An example of the first instance would have a patrol officer who is driving down the street see a suspect run out of a bank with a gun in one hand and a bag of money in the other. Just as this takes place, the officer receives a radio call describing a bank robbery at that location. Another example would be when an officer arrives at the scene of a call and sees a dead body with two gunshot wounds to the back of the head. A second person at the site shows the officer a gun and says that he saw someone shoot the person lying on the ground and then run away.

Before the Crime Has Occurred

In the second instance, an investigator may have an informant who has reported that a burglary is going to occur. The informant tells the officer the names of some of the suspects but does not know the location or exact time the burglary is going to take place. The officer would very likely begin the investigation with an attempt to identify the suspected parties and to figure out where the crime is going to take place. These beginning steps are the start of an investigation. Another example of the second instance might occur if a department had several reports of stolen cars from a shopping mall and set up a surveillance to catch those responsible. One day an officer on a stakeout sees a person walk up to a parked car, break the window, and after modifying the ignition, drive away in the car.

ADMINISTRATIVE INVESTIGATIONS

The parameters of an **administrative** or noncriminal **investigation** are broader than those of a criminal matter. If the purpose of the investigation is not illegal and the civil rights of those involved are not infringed on, an administrative or civil investigation may proceed. Administrative investigations are used in both the public and private sectors and most commonly occur when the intent is to resolve a matter without a criminal prosecution. In the public sector, many agencies have department policies, general orders, or even written management practices that cover all aspects of the types of conduct that should result in an administrative investigation, how a complaint should be received and documented, who should conduct an investigation, and how the report should be formatted, and what should be the basis of the final disposition. Some behaviors that might cause a law enforcement agency to initiate an administrative investigation include rudeness, discourtesy, insubordination, neglect of duty, and the all-inclusive conduct unbecoming an officer. Examples of administrative investigations might include a review of property tax records to find out who owns a particular piece of land or how many square feet a building contains. The information contained in public records needed to complete any of these investigations is not only open to review but is also available for use.

In the private sector, the proprietary rights of an employer are often the basis for an administrative inquiry. Employers have an obligation to provide a safe and secure workplace for their employees. One of the ways they accomplish this is through work rules that clearly set standards of behavior and control access to the workplace. Generally, an employer in the private sector has the right to use physical and electronic measures to monitor and control what comes into their work site. When information is discovered through the use of these security tools and it appears that an employee has violated local work rules, there is often justification for the employer to initiate an administrative investigation. Administrative investigations also might begin because an employee is found to possess some kind of contraband or if there is an indication that an employee is falsely claiming entitlement to benefits. They might also include a look into whether or not there has been a misappropriation of company products or supplies or if someone is creating a hostile environment in the workplace.

How and why administrative investigations begin often mirrors the way criminal matters are undertaken. In an effort to strengthen workplace safety, many companies conduct safety and security training for all of their employees. This training usually includes information about how employees can report suspicious or improper behavior. In addition, many locations conduct random inspections of employees as they arrive at work or leave for the day to ensure that contraband is not being brought into a facility and that company property is not being removed. A third way that administrative investigations may be initiated is through information discovered during routine audits of financial records or electronic files.

Regardless of which occurs first—the administrative investigation or the reason for it—investigators must constantly be aware of all aspects of the environment. This means they have to monitor and adjust their methods to all conditions, circumstances, and influences in the investigation. Are the actions they are taking going to impede the ongoing purpose of the company the investigators are working for? Are the interviews they are planning to do going to disrupt the workforce or cause production to stop? If so, it may be appropriate for the investigators to meet with the business managers responsible for the employer's operations and resolve the possible negative outcomes before proceeding.

Conducting the investigation appropriately and handling problems or issues at this point make the report writing part of the job a lot easier.

Investigators must also recognize that persons who appear to be uninvolved may in reality be suspects or witnesses. They may have something to gain or lose from disclosing or withholding their involvement in the matter. The people you are talking to may not honor your request to keep the matter confidential. It is human-nature to talk to friends and co-workers about the events in our lives. Investigators should expect some of the people they talk with to tell others about the investigation and even about their conversations.

The investigator must also be aware that evidence may be difficult to locate, incomplete, or latent at first glance. In order to make the most of every opportunity to preserve evidence and to ensure a complete investigation, the investigator should always strive to involve evidence specialists at the earliest time. Advances in technology are happening all the time, and some of them make it possible to get useful information from evidence that was previously worked and found to be of no value.

THE STEPS IN INITIATING AN INVESTIGATION

Depending on the circumstances and regardless of whether an inquiry is a criminal or administrative one, there are some basic steps that all investigators can take to initiate an investigation. As each opportunity to investigate begins, the good investigator considers these guidelines.

1. Arrive at the place you are needed to begin the investigation. More than one serious crime has gone without an investigation beginning promptly because someone crucial to the case became sidetracked or unavailable while getting to the scene. Crashing on the way or stopping to write a minor traffic citation is generally considered bad form.
2. There may be people at the scene who are in need of help or assistance. Remember that in emergency responses, the first order of business is life safety, and a big part of any investigator's job is to help those in need. Helping someone in need can go a long way toward developing a spirit of cooperation between the people with the information and the person who needs it—YOU.
3. If the circumstances dictate, you may face a suspect who must be detained or arrested even though minimal evidence is available. If this is the case, and the law allows the suspect's arrest, it should be done to prevent an escape and eventual hunt to locate him or her, which would threaten the case. This is not to say that you should arrest someone without cause; that is something that should never be done. But you

should be prepared to move forward quickly if circumstances so dictate.

4. Find people with information about the case as soon as possible and talk to them right away. This may prove challenging if there are a number of persons who witnessed the event and have information to share with you. Remember that people who witness a crime may not want to share their information with you. They may not be in a position to talk to the police in front of their friends or associates for fear of retribution. They may have alternate plans that require them to leave the area right away, which makes your time frame for interviewing them unsuitable to their situation. Establishing a way to get in touch with them at a later time with more suitable circumstances is a good thing to do.

5. Protect any physical evidence by setting up some kind of control or security around the crime scene. Although it may not look like there is any evidence of value during your preliminary check, there may be important clues that a trained evidence technician will be able to locate, process, and testify about in a trial or administrative hearing. It is also important to realize that even though you do not think there is evidence at the scene, every step you take as you walk through the location may be the one that destroys what evidence there was. Remember that once evidence is gone, it can rarely be replaced. Having solid and consistent habits in dealing with evidence will benefit you throughout your career. Whether you are outside at an industrial site or inside a bank, taking care of the evidence starts the moment you arrive. Take a good look around and start right off by doing the right thing. This will allow you to establish and maintain a solid chain of custody of the evidence in your cases.

6. Interviewing a suspect is a good way to get information about what happened, but this is an area greatly determined by the circumstances of the situation, not the least of which is the suspect's attitude and his or her level of cooperation. It may be advisable to wait until a full assessment of the case at hand can be made before doing any suspect interviews. A good rule of thumb is that if you are in doubt, always get a second opinion from an experienced investigator or a representative from the district attorney's office before proceeding. Knowing all you can about the situation and person you are going to talk to can only help you. There is no substitute for good preparation.

7. No one can remember everything, and now is the time to help your memory by starting to write your investigative notes. Good notes will be invaluable to you when you begin writing the report. Knowing how to take quality notes is a practiced art. Having the right notebook for the job is the first step, but using it successfully is a key to being a good investigator.

8. Crime scene investigators do not just appear out of thin air. Someone must notify them that they are needed at the crime scene. If you are in charge of the investigation, the responsibility to make the notification is yours. Evidence specialists are good at their jobs, so let them help you do yours. This can take some time. The size and complexity of a crime scene dictates how much time will be needed to process a scene. Evidence specialists will document their actions and give you a report. This part of the investigation should not be rushed. If you are in charge, let them have all the time they need. It will pay big dividends for you in the end.

9. Get some help. If you need someone with expertise in a certain area, do not hesitate—make the call. You can expect to run into crimes and crime scenes that are new and complex. Get an expert to help you.

10. Write the report.

How It Happens

In the ideal investigation, information comes to the investigator in a logical and understandable manner. In the real world, few if any investigations take place under ideal circumstances. You will probably agree that major investigations around the world take place in less than ideal circumstances. They can involve hundreds of investigators and very large crime scenes. This creates the need to develop information and store it in a systematic way that is accessible by those who need it. It is important for an investigator to recognize that information gathering is the key to solving cases.

Average or Superior

The **average investigator** looks at a group of people near a crime scene and says, "There is a group of people standing around my crime scene." The **superior investigator** looks at the same group and says, "There is a group of people standing around my crime scene and each of them knows something about this case. It's my job to get them to tell me what they know and then figure out how to get the ones who don't want to talk to me to do so anyway."

THE WHO, WHAT, WHERE, WHEN, WHY, AND HOW OF INVESTIGATIONS

Some of the things you may be able to find out about the case include who is involved and what their backgrounds are, where they are and where they should be, who their associates are, and a lot of information about the crime, especially the WHO, WHAT, WHERE, WHEN, WHY, and HOW. Many people use these six words, also referred to as the five Ws and the H of investigation, to describe what they are looking for. However, without some explanation or example, they might remain just words. When good investigators use these words, some of the things they think of are:

Who

Who is the reporting party?

Who is involved as the victim?

Who is the suspect?

Who are the suspect's friends and associates?

Who knows what happened?

Whom do I need to talk to?

Who is a witness?

Who was with the victim?

What

What happened?

What was the victim doing?

What was used to commit the crime?

What has happened since I was called?

What do we know so far?

What are the next five things that need to be done?

What time did it happen?

What time was it discovered?

What time was I called?

What was the relationship between the suspect and victim?

Where

Where did it happen?

Where is the victim?

Where is the informant?

Where is the evidence?

Where were the witnesses?

When

When did it happen?

When was it reported?

When was the victim last seen?

When was the suspect last seen?

Why

Why did it happen?

Why was it reported?

Why did witnesses not tell you something?

Why did witnesses tell you certain things?

Why did the crime happen the way it did?

How

How did the event happen?

How did the suspect get there?

How did the victim get there?

How did the suspect get away?

How did the suspect know the victim?

How was the event reported?

How old is the crime?

How much more needs to be done?

There is a lot to be said for the experienced investigator who has seen or done it all a hundred times over, knows all the tricks in the book, and knows when to use the magic that accompanies this knowledge. But much success has also come to those who, although not as experienced, are willing to work hard and learn from the actions of others. One should never rule out luck, or a hunch, as a way to turn a dead-end inquiry into a viable and active investigation. This is not downplaying the abilities of the thousands of superior investigators who are very successful at their jobs, but it is to make a point that the uncontrollable facts of daily life may turn out to be a blessing in disguise.

If there were a way to list the characteristics and traits that superior investigators possess and have passed on to one another, it would surely require more than a few pages to do so. There does, however, appear to be some common thread woven through the cloth of the superior investigator that holds together these qualities, which indicate a greater chance for success.

THE QUALITIES OF A SUPERIOR INVESTIGATOR

1. **Superior investigators are ethical.** They inherently know the difference between right and wrong behavior as well as what is necessary for them to be considered a professional. They play by the rules and do not get involved in the manufacture of false evidence or testimony.
2. **Superior investigators are aware.** They are aware of their surroundings and are able to quickly evaluate situations and circumstances, as well as people, and are accurate in their assessments. They have a broad knowledge of local and world events, which gives them the ability to place occurrences in perspective and forecast strategies and alternate courses of action with a high probability of success.
3. **Superior investigators are energized.** The long and tedious hours of fact gathering and analysis do not overwhelm them. The day is never too long to finish some aspect of the case. Even when all seems lost and there are no prospects for success within grasp, the superior investigator can reach

within and find the energy and endurance to take the next step, to do one more interview, to read one more supplemental report, or to make one more phone call on the chance it might turn into a workable lead. They give 100 percent all of the time and expect everyone else to do the same.

4. **Superior investigators think outside the box.** They have the ability to see beyond the next two or three steps in the investigative process and, thus, are able to visualize new or unusual ideas or techniques that may prove to be helpful in solving the case. They are not afraid to try something new or unusual for the sake of achieving success in the case. They constantly evaluate the successful ideas and tasks, which they and others routinely perform in a process of continuous improvement.

5. **Superior investigators are determined.** They have the ability to move forward, even in the face of the seemingly unachievable. When others might hang it up and say, "Enough is enough," superior investigators are saying, "Well, so far I haven't been successful, but I've tried only a few ideas." Superior investigators never give up. They may prioritize and reprioritize their cases, and put the ones with few or no workable leads on the back burner, but they are always thinking about what they need to do to find the truth.

Just as the job of a good investigator has been described, it is equally important to describe what an investigator's job is not. It is not the investigator's job to prove someone is guilty or that a particular person did not commit the crime. It is the investigator's job to report all information, regardless of the impact the fact pattern may have on the case. An investigation must be unbiased and impartial to be valid. An investigator gains **credibility** by doing a thorough and professional job on all inquiries undertaken. It is not necessary or realistic for an investigator to find a solution for every investigation he or she starts, nor is such success necessary to be considered a superior investigator or to have credibility. Superior investigators do all of the little things right the first time, allowing them to start the big things on a positive and timely note. This creates the opportunity for success and allows for the best possible outcome. This in turn allows them to establish credibility, something that is hard to come by yet easy to lose.

SUMMARY

Whether an investigation is a criminal inquiry or an administrative matter, an investigator needs to have a lawful reason to conduct one. Investigators should not try to prove anything; rather, they should try to find the truth and let the facts prove or disprove any allegations. Investigators should always strive to work in an orderly manner and try to logically follow a game plan that covers everything from getting to the crime scene to writing the report. One of the things that can help an investigator write a good report is knowing the who, what, where, when, why, and how of what happened.

It is not an easy task to solve the complex investigations that confront investigators today. Qualities that can help investigators achieve success include being ethical and always doing the right thing, being aware of their surroundings and having the ability to size things up, being able to work long and hard when necessary, being willing to try new approaches or techniques when necessary, and being able to prioritize their cases so they can devote their energy and talent to those cases with the best chance of being solved. With these qualities at work and always remembering that doing a thorough and professional job on every case shapes their credibility, investigators will enjoy success.

REVIEW

1. An investigation is a lawful search for things or people.
2. The goal of an investigation is to find the truth.
3. Investigations can start before or after the crime.
4. Protect the crime scene.
5. Establish a chain of custody for evidence.
6. Locate witnesses as soon as possible.
7. Involve experts in evidence handling.
8. Report all information.
9. Use the who, what, where, when, why, and how to get started.
10. Good investigators are aware, energized, determined, and think outside the box.

EXERCISES

1. Match the example of superior traits with the quality

   a. Perceptive   d. Creative
   b. Tireless   e. Persistent
   c. Ethical

   _____ Reviews old cases, looking for something new to try.

   _____ Interviews a witness three or four times.

   _____ Tries a new undercover operation.

   _____ Connects names and events in the news to an old case.

   _____ Reprioritizes what needs to be done on a case.

   _____ Is willing to work long hours to get the job done.

_____ Always does the legally and morally correct thing.

_____ Never fabricates evidence.

_____ Uses a sting operation to catch a crook.

_____ Uses events to forecast likely outcomes and scenarios.

2. Match the WHO, WHAT, WHERE, WHEN, WHY, and HOW with the example. Answers may vary.

   a. Who   d. When
   b. What   e. Why
   c. Where   f. How

_____ The television was worth $700.

_____ Dave was stabbed at 1200 hours.

_____ The door was kicked in.

_____ Ron Ho called at 1845 hours, reporting gunshots in the area.

_____ Joe Smith was arrested at 6th and Main.

_____ Footprints were found outside the broken window.

_____ The silent alarm was received at 2100 hours.

_____ Pete Brown was shot in the back of the neck.

_____ Fava saw a red car drive away.

_____ The suspects took the necklace from the display case.

_____ The knife was marked KCB and booked at central property.

_____ Najjar opened the store and saw the broken window.

_____ Just before the shooting, Brown was drinking at the bar.

_____ Bill Smith was lying on the bathroom floor.

_____ Dispatch received the first call at 0630 hours.

_____ The coroner identified the victim as Ron Lincoln.

_____ Detective Colin just left the crime scene.

_____ The body was found under the bridge.

_____ The victim's wife purchased a $300,000 insurance policy last week.

_____ I found a stolen bicycle in the bushes.

3. Visit a library or bookstore and review several books about real-life investigations. Select two books of interest to you and while reading them, try to determine:

   a. What was the reason for the investigation?
   b. Are there examples of the investigator following or not following the guidelines discussed in Chapter 2?
   c. What traits of the superior investigator are evident?
   d. In your opinion, which trait was most important to the successful completion of the investigation?

4. Identify a criminal investigation in which the lead detective did not exhibit ethical behavior. What are the consequences for the involved investigator, his or her agency, and the profession?

5. Read about reported investigations in the newspaper for one week. Brainstorm the investigative techniques used in the cases. Can you suggest others that might be successful?

6. Working with three or four classmates, construct a bibliography of investigative books. Use this bibliography to develop your knowledge of investigative techniques.

QUIZ

1. The definition of an investigation is:

   a. An inquiry by a professional
   b. A search for a stolen television
   c. A lawful search for a thing or person
   d. An undercover search of a campaign office

2. The goal of an investigation is to:

   a. Find out what happened
   b. Find out what was taken
   c. Get to the bottom of the matter
   d. Find the truth

3. Generally speaking, the less evidence you have, the better your case will be.

   a. True
   b. False

4. When you begin an investigation, your first consideration should be:

   a. Catching the suspect
   b. Getting there
   c. Protecting the evidence
   d. Locating the best witness
   e. Making an arrest

5. The best person to call to take care of the evidence at a crime scene is:

   a. A sergeant with 20 years of experience
   b. Your partner
   c. A crime scene investigator
   d. A volunteer who has completed an evidence class

6. Good investigators should report:

   a. All information
   b. Only the information that helps show guilt

    c. About 75 percent of the information they discovered
    d. Only the information that shows innocence

7. Which of the following is *not* a public record?

    a. The number of arrests a person has
    b. How much taxes a property owner pays
    c. The number of civil actions a person has been a plaintiff in
    d. The number of convictions a person has

8. To what do the five Ws and the H of investigation refer?

9. What are the two main ways an investigation can begin?

10. Name the qualities of superior investigators.

3

Note Taking

KEY POINTS

Notes, field notes, investigative notes—regardless of which term you use to describe the initial information gathering that good investigators do, you will need to have good notes in order to write a good report. This chapter will combine theory with practicality by discussing not only the importance and basic uses of field notes but also the mechanics of note taking, types of field notebooks, and how to organize a field notebook. We will also look at the pros and cons of using a tape recorder to accomplish the same task that you could using paper and pencil.

KEY TERMS

Accurate	Measurements
Aid to memory	Mechanics of note taking
Building blocks	Readable
Concise	Sketch
Drawing	Storage
Key words and phrases	Thirty-word version of a story
Listen first, then write	

THE IMPORTANCE OF FIELD NOTES

Few people have the ability to remember everything they do, see, or hear. The short-term recall most people have is poor at best and virtually nonexistent at worst. In most cases, the longer the time span between the occurrence of an event and the reporting of it, the greater the chance that an incomplete version of the story will be told. Further, as time passes, it is more likely that inaccurate facts or mistakes will be presented and that something important will be left out of the story. Think back for a few minutes and try to remember what you had for breakfast today, what color shirt you wore yesterday, or who were the first three people you talked to two days ago. You were probably able to remember because they are recent events to which you are personally connected. But try to recall the same information from three weeks ago and you will see that it is much more difficult. I think you will agree that the longer it is between something happening and trying to recall it, the more unlikely the latter becomes. Also, consider the example presented by a college class. We all know that there will be a final exam at the end of the semester and quite possibly a midterm exam partway through the course. We also know and recognize the value of having complete and accurate notes in order to study for the exam. Even with this knowledge and the opportunity to take notes and compare them with our classmates, we find ourselves at the end of the semester looking at an exam that contains questions we knew would be on it and having no idea how to answer them.

With regard to investigative report writing, we know that the events we are investigating are going to result in questions at a later time. It may be a few hours, days, or weeks—or in some cases several months or years in coming—but nonetheless the questions will come. Fortunately, we have the opportunity to do something about this during the investigation that will help us when the exam—usually in the form of a trial—may require our testimony. This something is note taking, and it is an essential part of the investigative report writing process. It would be a wonderful world if our memories were so good we never had to refresh them, but that is not the case. You will need notes, and you will need to refresh your memory. Fortunately, there are a few basic rules and guidelines that will make the note taking function of report writing much easier to understand and accomplish. Inasmuch as we have discussed why we need to take notes, it may also be helpful to understand some of the basic uses for them.

BASIC USES OF FIELD NOTES

Storage

The old adage "A place for everything and everything in its place" applies here. Both public and private record keeping systems, as efficient as they are, are not perfect enterprises. Over the course of many months and years, documents are misplaced, lost, or destroyed. Whether these events occur by accident or with deliberateness, the end result is that an official document is no longer available. The responsibility for replacing the lost or unavailable documents rests with the person who originally completed the report. Although this may seem unfair to some, it really is the best method to replace whatever is missing. No one else has the knowledge of what the report may contain. In addition, no one else has the original notes that would allow the investigator to talk to the involved parties and once again find pertinent information. In their most basic form, an investigator's notes are the first and last level of record **storage.**

Building Blocks

A report just does not appear from thin air. Much work and thought goes into writing a thorough and complete investigative report. It is of paramount importance that an investigator has written notes to refer to as he or she completes the report. An analogy can be drawn between the investigative process and report writing tasks to certain parts of a construction project. Investigative note taking and report writing are much like preparing a building site for new construction. A foundation must be completed before the walls can be built. Notes are the foundation and the **building blocks,** and the report that follows is like the building that rises from this first important step. Your notes are the most complete supply of raw material for a report. If you have done your job, you should have information from several people involved in the case—descriptions of suspects, property, and evidence—as well as where it was found. In short, you should have what is needed to construct a report and continue working the case.

Aid to Memory

Investigators with photographic memories may never need to take notes as an **aid to** their **memory** during their investigations because they will be able to remember all of the details. Most investigators do not have photographic memories, however, and do require good notes to remember not only the things to include in the report but also the unreportable facts. What seems trivial at the earliest stages of an investigation may be very important several months later when an unknown witness or recently discovered evidence surfaces. A fairly common occurrence in recent years is for convicted criminals to confess to serial crimes that they committed years ago. In order to prove that they did the crimes, they offer small details of the crime scenes or victim descriptions that could be known only by someone at the scene. Where is the best place to look for this information? The answer is in the notes taken by the investigating officers. These notes can help officers remember details that no one else would and also help in formulating questions that could eliminate a publicity seeker who is making a false confession. Complete notes are invaluable in these circumstances.

THE MECHANICS OF NOTE TAKING

The **mechanics of note taking** refers to the practical application of writing down the initial findings of an investigation in a notebook. It includes the selection of a specific type of notebook and an appropriate writing instrument. The mechanics of note taking are simple to master with practice. It is desirable to take notes in a way that allows anyone who reads them to understand and interpret what they mean. For example, some individuals have handwriting that is difficult to read. This presents a problem if a case is going to involve more than one investigator or requires several people to review and share recorded field notes.

Readable

There are few rules on the type of handwriting that should be used to record field notes. Whether you print or write in cursive, the end result is what is important. That end result is that your notes need to be **readable** and understandable when they are needed down the road. Good penmanship is a learnable skill and the adage "practice makes perfect" certainly applies here.

Accurate

It is absolutely critical that you take **accurate** notes in the field. This means that the statements you write are correct, measurements are precise, names are spelled correctly, and phone numbers and addresses are without error.

Factual

Notes not only need to be accurate but also to be factual. Facts are things that can be proven. It is important to keep personal opinions out of a notebook. Experience will help you learn what to write in a notebook and what to leave out, but try to write only facts and keep your ideas about how something happened out of the notebook.

Concise

Whenever you are asked to write a lot of information in a small space, the tendency is to use symbols, codes, or abbreviations to get the job done. Although the use of this type of writing is permissible, it should be avoided as much as possible because different marks and symbols can mean different things to different people. You must also remember that the secret codes or abbreviations you are using today may not mean the same thing to you several months from now. Instead, be **concise** and get the most out of as few words as possible.

Few things will prove as embarrassing as not being able to read and understand your own notes. Avoid this problem by using clear, complete words whenever possible. The goal of note taking is to record as much information as possible about an investigation in a concise, readable, and understandable manner. Develop good habits and improve on them as you gain experience. It will pay big dividends for you over the length of your career.

TYPES OF NOTEBOOKS

There are many types of notebooks that may be used to record field notes. The key here is that the investigator should feel comfortable with the notebook he or she is using, and that it allows for easy writing in a variety of situations. There will be many instances, such as during a surveillance, where the investigator will have to take notes while standing, sitting, walking, or even lying down. It may be necessary to record notes during the day or night and whether it is sunny outside or raining. As you can see, the notebook must be a versatile tool.

In all probability there is no single notebook that meets everyone's needs. Smaller sizes such as a 3" by 5" pocket notebook allow for easy storage because they will fit into a shirt or pants pocket. The drawback is that with such small pages it is difficult to record much information. Something like a 6" by 9" spiral-bound steno notebook has the advantage of being big enough to write necessary information in it, yet is still manageable enough to easily hold and be stored in a pants pocket or inside the belt line at the small of the back. A full-size pad of paper gives the note taker ample room to write but may be awkward to hold and store. Another disadvantage is the ease with which pages may become detached from a large tablet of paper.

An investigator should try several sizes and styles of notebooks until he or she finds the one that best meets his or her needs. A suggestion is to hold each type of notebook under consideration and, while standing in an area dark enough to require the use of a flashlight, write enough information to fill five pages. This should give you an idea of the versatility of the notebook and whether or not it will meet your needs. If the notebook passes this test, it will likely allow for easy use in any situation or environment an investigator may be involved in, such as in a car, in a house, or in inclement weather. How you organize your field notebook can have an impact on how successful you are in using it. Filling a notebook with information is only part of the battle. You must be able to access the information and retrieve what you need when you need it.

ACCESSING THE INFORMATION

Taking complete notes is certainly a goal in the report writing process, but it is not the only thing that you should be thinking about. Just as important as having complete information is being able to go back to it a few hours later—or, for that matter, days, weeks, or months later—when you are writing the report. This is why the habits you use in setting up your notebook are important.

It is generally recommended that you write on only one side of the page. When you use both the front and back sides, the notebook can become difficult to handle and manipulate while turning pages and adding information. Use a writing instrument that will not bleed through the page or smear if touched. A good mechanical pencil or ballpoint pen might be a good choice for field officers. Both will give a quality impression on the page and will work in most circumstances. Whatever writing instrument you select, make sure it will stand the test of time and not bleed through the page. A neat, clean, and well-organized notebook is a great tool for the investigator to have.

Organizing the layout of a notebook page is another detail that can have a big impact on how useful this tool is for you. One method is to establish a clean break between dates by consistently closing out the bottom of the page by placing some type of ending graphic drawn across the page right after the last entry. Start the next date at the top of a page and begin by writing the date, shift, area, car assigned, and partner's name and serial number if applicable. Create a margin along the left side of the page that can serve as an index to the material. Entries in the margin such as ROBBERY would be followed on the page with the details of the crime or suspect description. Most officers start the day in roll call and receive information about crimes that have happened during the past 24 hours. Indexing these crimes in the left margin will allow you to refer to them quickly when needed. Once you go into the field, add information to your notebook in the chronological order you receive it.

Leave Some Room

One tendency many new officers have is to write too much on each page. Get in the habit of not crowding too much information together. Leave some space on each line and skip a line or two between different incidents or entries. When the page is set up properly, it will be pleasing to the eye and allow you to glance at it and quickly find what you are looking for. A notebook is one of the least expensive items you will need in your career. Most agencies supply notebooks to field personnel, but even if you have to buy your own, remember that no cop ever went broke buying a notebook. Leave some room, keep your entries neat, and let the notebook work for you.

WHAT SHOULD BE INCLUDED

Experience is a great teacher in determining what should be included in the note taking process, but there are several items of information that are important to include in every case. Accurate names, addresses, and phone numbers top the list of important information. The importance of recording this type of information accurately cannot be overly stressed. Incorrectly identifying information results in wasted time and an incomplete and sloppy investigation.

Weather

There may be occasions when the weather conditions will have an impact on a case you are investigating. If necessary, write the important items such as temperature, whether or not it is raining or has recently rained, and wind conditions in your field notes. Other important information includes measurements, sketches, drawings, and key words or phrases used by suspects or witnesses.

Measurements

As simple as it may seem, making and including accurate **measurements** in your field notes can sometimes be a challenge. Fortunately, a little preparation and a plan are usually all that is needed. Many investigators carry a tape measure with them for just such occasions and are able to record very accurate information for shorter distances. Experienced investigators have also learned how to step off longer distances and record an accurate measurement; they have measured their stride and know what a three-foot step is.

The second half of the preparation is to know how to measure. You should always find two permanent fixed objects to measure from in order to have an accurate set of measurements. Using objects such as painted lines, automobiles, or temporary signs can result in inaccurate or questionable measurements. Instead, try finding telephone poles, concrete curb lines, building foundations, or surveyor's marks. Then measure from two of these, in different directions, to come up with the location you are describing. Always include a note that your measurements are approximate and whether you used a tape or the step method.

Sketches

Including a crime scene **sketch** in your field notes may be appropriate in a number of cases. If you do choose to draw one, take your time and make it as complete as possible. Use fixed objects as points of reference for your measurements and try to keep the drawing proportional. Today, a video or digital camera may be available, but there is nothing wrong with supplementing the images made by these cameras. If your notebook is not big enough to draw an accurate or complete picture, use a full-size sheet of paper, include the case number and the date and time you made the drawing, and be sure your name is on it. The fact that you completed a crime scene sketch should be included in your field notes.

Drawings

Sometimes it may be necessary to draw a picture of an object or a special mark on an item for identification. Descriptions of jewelry seem to provide opportunities to use this skill. Allowing a victim or witness the opportunity to make a **drawing** of a ring or pendant can go a long way toward identifying it later.

Some people also take great pride in putting their initials or personal mark on their belongings. Tools, toys, fishing equipment, and bicycles are some of the things that people identify with a unique mark. Having a drawing of this mark will not only help establish ownership but also can become probable cause when a cop on the beat finds a car full of tools with a unique mark that was discussed and shown in roll call that day. The adage "a picture is worth a thousand words" is proved true time and time again by this technique.

Key Words and Phrases

There are only a couple of times you will need to show the **key words and phrases** someone used. One is when you are writing the words the suspect used during the commission of a crime, and the other is when a suspect is confessing to a crime.

In the first instance, a witness is usually going to tell you the exact words he or she heard the suspect say. It is important to accurately record these because this might be part of the suspects M.O., or method of operation. The second instance in which the exact words are extremely important is when the suspect is confessing to a crime. It is good to have the exact words he or she said during the admission. During these two occasions, take your time and get it right.

Investigative notes need to be complete enough to help you write an accurate report but not so in-depth that they are as long as the finished product you ultimately produce.

Personal Information

Last but not least, the notebook should contain the name, business address, and phone number of the investigator along with the inclusive dates the information in the notebook covers. This identifying information should be clearly printed on the front cover.

Remember that this is a business notebook and you are a professional; therefore, the notebook should contain only business-related information. Always take care to leave personal information out of the notebook because it has no place in a professional working document and may be subject to the power of subpoena.

How an investigator takes field notes will depend a great deal on the situation and how comfortable the person being interviewed is about sharing the information she or he knows. Some ability to write quickly and accurately is needed, and these skills can be improved with practice.

As the investigator begins the interview, his or her manner and demeanor are just as important as the authority or credentials he or she carries, if not more so. The professional investigator will seek the help of a witness rather than expect it and will convey that attitude in all dealings with those whom he or she encounters. A proven method for gaining information in any interview is to start by identifying and introducing yourself to the person you are about to interview and then ask him or her to tell you a little bit about what happened. If the person agrees to talk with you and expresses any cooperation, you should let him or her talk without interruption and without writing anything down. This will give you the opportunity to evaluate the person's story without being distracted by writing while he or she is talking. To expedite the interview process, ask the person to tell you what happened in thirty to forty words. This is sometimes referred to as the **"thirty-word version" of the story.** While they are talking, you may be able to formulate specific questions based on the information the person does and does not tell you. Hearing this short version may also help you identify the type of crime or incident involved. Many people will come home, find that someone has broken into their house and stolen something, and report the event to the police by saying, "I've been robbed." It is easily understood why the victim of such a crime would report it as a robbery when it is really a burglary. However, giving the victim the chance to tell you about the crime in her or his own words gives you the opportunity to formulate questions based on the overview you hear. Also, it gives you the opportunity to determine what happened. Understanding the type of crime is important to the investigator, who may need to locate specific evidence or information to establish the corpus, or elements, of the crime.

Once the witness has given you the "thirty-word version," you should ask specific questions that require specific answers. This will provide you the basic information needed in order to complete the crime or incident report. You should have a thorough knowledge of the format of the crime report forms you are using and should use this format as a guide to formulate and sequence your questions. By doing so, you will be able to write the information into your notebook in the same order it will be needed to complete the report. Putting this information into chronological order will save you time and effort in preparing the finished product. This is only a suggestion and should never be used as a rigid rule that would limit the amount or quality of information you could gain from someone. Common sense is a great asset for the professional investigator and will almost always allow for a successful investigation.

Once you have completed the task of writing down the identifying information, ask the witness to tell you what happened and try to write down ideas or major thoughts as they are presented to you. After you have completed the question and answer session with the witness, it is usually a good idea to review the information with him or her and give the person the opportunity to add and change anything that he or she has misstated to you. Remember that it is your obligation to write an accurate and unbiased report. Giving the victim or witness the chance to alter or correct his or her story at this time helps to ensure that you do your job. It never hurts to conclude the interview by asking if there is anything else the person wants to add and by giving him or her a phone number at which to call you if he or she needs to speak to you again.

RECORDING NOTES

Using handwritten notes is a solid basic step many investigators use to collect information about their cases, but it is not the only method available. Some investigators use handheld tape recorders as an additional source, and in some cases, as a primary source. The handwritten style might be best for beginning investigators because all of the information collected can be reviewed at a glance. A tape recorder gives more seasoned investigators, the opportunity to collect a lot of information in a short time. Although this can be a huge asset if the crime scene is larger and involved, it can also have drawbacks.

The tapes must be transcribed and the transcriptions certified as accurate, and then they must be preserved just as handwritten notes would be. Discrepancies between the recorded notes and the final transcribed version will always be an issue in a trial. Other potential problems with using a tape recorder can include loss of power, mechanical problems, poor tape quality, and loud background noise. The opportunity to use a tape recorder to gather field notes may prove to be a timesaver and an aid to many officers. However, even with departmental policy allowing the use of a recorder, the individual investigator must make sure it is right for him- or herself.

Regardless of whether you are handwriting notes or using a handheld tape recorder to capture the information, the two most important things to remember and practice with regard to note taking for investigative report writing purposes are to **listen first, then write,** and to make sure you can understand your notes.

SUMMARY

Just as the beginning of an investigation is defined by when and how the investigator gets involved, the roots of a report are the field or investigative notes the investigator compiles. Notes are so important because humans are usually ill equipped to accurately remember details over extended periods of time. Good notes are a great aid to the investigator and serve as a place to store information, as building blocks for reports and as an aid to the investigator's memory.

To be of the most value, notes should be readable, accurate, and concise, and the investigator must always be able to understand what has been written. Investigators should give careful thought to the type and size of field notebook they use because a good fit will only make the job easier. Once the investigator has chosen the best style and size of notebook, a plan should be used in setting up how the information will be recorded and then accessed. Investigators cannot go wrong by leaving some room between their writings, which will make reading and locating information easier. Critical items such as weather conditions, measurements, sketches, drawings, and key words and phrases used by the suspect should be carefully added. Just as important as it is to include this type of information, it is equally important to leave personal information out.

Knowing how to gather information is another skill that marks the superior investigator. Asking a person to give a short overview of what he or she knows, before you start writing can be of great assistance in getting complete and accurate data because it can help you formulate the right questions to ask as you get more in-depth. Last, even though handwritten notes are more commonly taken, some investigators will use a handheld tape recorder to capture their notes. Although this can be a quality tool for some investigators, it is not without problems for others. Loss of power, mechanical problems, bad tape quality, and background noise are some of the problems investigators may face. Whichever method is used, notes must be cared for and retained according to departmental policy and case law.

REVIEW

1. Basic uses of notes are:

    Storage of information

    Building blocks for reports

    An aid to your memory

2. Notes must be legible and understandable.

3. Information must be accurate.

4. The type and style of notebook is an individual matter.

5. Keep personal and nonbusiness information out of your notes.

6. Remember to ask for the thirty-word version.

7. Listen first, then write.

8. Make sure you can understand your notes.

9. Write on only one side of the page.

EXERCISES

1. Review your class notes from any college course or training session you have completed and consider whether or not:

   a. Your notes are an aid to your memory about what took place in the class.
   b. Your notes successfully serve as a storage place for the information you need.
   c. You are able to prepare a one-page report about the class using the notes as a building block for the report.
   d. There is any personal information in the notes.

2. Record a thirty-minute news broadcast, and watch it while recording it. Watch it a second time while taking notes, and use these notes to prepare a report about the stories that were broadcast. Which principle of note taking will this help you practice?

3. Repeat exercise 2 with a different tape, but do so while standing in a dark room, using only a flashlight and the light from a television screen to illuminate your notebook. Is there any difference in the mechanics of note taking?

4. Visit a stationery store and look at several types of notebooks. Make a list of the pros and cons of each notebook you examined. Which type and size do you prefer?

5. Interview a classmate and record enough information so that you can introduce the person to the class. Pay particular attention to the correct spelling of names.

6. Watch a television program with a "cops-and-robbers" theme. Record the information you think is important, and then prepare a one-page report from your notes.

7. Review a classmate's notebook. How is it organized? Are the notes legible and understandable?

8. Use a field notebook to sketch a diagram of your classroom and include any audiovisual equipment in the room. Be sure to include measurements of the room and the placement of the equipment.

9. Visit the campus library and sketch the first floor, including the measurements of all important equipment and tables.

QUIZ

1. Why is it important to have field notes?

   a. Recording all information is required by law.
   b. Few people can remember everything.

c. Taking notes gives an officer a way to look professional.

d. Notes offer the proof that something happened.

2. Using abbreviations in notes is a good practice to follow whenever possible.

a. True

b. False

3. The preferred type of handwriting for your field notes is:

a. Shorthand

b. Cursive

c. Block printing

d. Anything that is legible and understandable

4. What is the best size of field notebook?

a. The smaller, the better

b. Anything with removable pages

c. Clipboard size, to get lots of information

d. Whatever works best for each officer

5. What information should be included at the start of each day's notes?

a. Your name, service date, and car assignment

b. The date, your name, area, and car assignment

c. The date, your partner's name, area, and assignment

d. The date, your partner's name, area, and car assignment

6. Where should your name and other identifying information be written in a field notebook?

a. On the cover

b. On the last page

c. On both the first and last pages

d. On the first page

7. The basic principle to be followed in taking notes is:

a. be polite.

b. write everything you can.

c. listen first, then write.

d. use as many abbreviations as possible.

8. List the three basic uses of field notes as discussed in this chapter.

9. Field notes have been compared to the construction of a building site. Explain.

10. What is meant by the expression "field notes are an aid to an officer's memory"?

4

The Rules of Narrative Writing

KEY POINTS

Knowing how to do something and being able to do it are often two separate things. This chapter will not only provide you with a simple and virtually foolproof report writing skill set but also help you understand why these techniques will work. This chapter discusses things such as how to begin the narrative of a report, how to choose simple and meaningful words, how to get the most out of the words you choose, the difference between facts and opinions and why police reports should contain facts instead of opinions, and how to refer to people in a report and demonstrates them with relevant examples. This chapter is the main entrée in the meal of report writing.

KEY TERMS

Abbreviations

Abstract words

Active voice

Chronological order

Concrete words

Direct quotes

Facts

First person

Indirect quotes

Jargon

Names and titles

Opinions

Past tense

Radio code

Rules of narrative writing

Spelling

Most of the time when someone shares a story about what happened, she or he includes what the person knows happened, what she or he thinks happened, and even what she or he wished would have happened. Included in this perception of what happened will be both **FACTS** and **OPINIONS**—and there is a big difference. **Facts** are things based on an actual occurrence, something that has an actual existence—in short, facts are things that can be proven. **Opinions** are beliefs, someone's view, and a person's best guess based on what he or she knows, and they may not always be accurate. In most cases when writing police reports, investigators should include the facts and leave the opinions out. We will address the occasions when opinions are properly included in police reports in a later chapter, but for now, stick to the facts and keep your opinions to yourself.

If you had the ability to place the rules and regulations of all the investigative agencies in this country side by side, you would undoubtedly see that there are thousands of differences in the way they do things. But just as there are differences, there are many similarities as well. It is a safe bet that each of these agencies wants its investigators to be courteous, fair, and professional, as well as to do a good job. It is also a good bet that each of these agencies wants and expects its investigators to document their investigative efforts by writing reports. One of the differences in the way these agencies do business is in how reports are completed. It is not uncommon to find different report forms and formats from agency to agency, nor is it uncommon to find different forms and formats within a single agency. One reason for this difference is that when investigative reporting is controlled by automation, such as word processing or computer-aided record keeping, these systems sometimes require reports to be in specific formats in order for an interface between the investigator's work and the automated system to occur.

Fortunately, this usually pertains to the face sheet, heading, and fill-in-the-blanks type of information. Normally, there are no hard and fast rules about how the narrative portion or body of the report should be written and because of this, some investigators are challenged when it is time to write a report. One method that works in all investigative writings—including crime, arrest, supplemental, incident, evidence, and information reports—is to use a continuous, free-flowing, narrative style of writing. With this style, no subheadings, sidebars, labels, or other text dividers are used.

Ideally, the investigator follows what I call the **rules of narrative writing** and writes in the first person; past tense; active voice; in chronological order beginning with the date, time, and how he or she got involved; and using short, clear, concise, and concrete words.

THE FIRST RULE OF NARRATIVE WRITING—FIRST PERSON

When writing a report, you should write in the **first person** and refer to yourself as "I" or "me." This clearly identifies you as being the person doing the investigation, and it is a clean and simple way of describing who you are. *I* is one of the shortest words in our language, looks good to those who read your work, and sounds good to those who hear your testimony. Never use the term "the undersigned" or "U/S" to refer to yourself. As the writer of an investigative report, you are always correct in using the personal pronouns "I" and "me" when referring to yourself. For example:

> I saw the pry marks on the windowsill.
> I interviewed Rand.
> Colin gave me the pistol.
> I found the marijuana.
> Najjar and I booked Edwards.
> Ho told me he searched Lewis.

THE SECOND RULE OF NARRATIVE WRITING—PAST TENSE

The events you are writing about have already happened and as such are part of history. Inasmuch as they have already taken place, it is quite proper to write about them as past events. The verb tense is the part of grammar that tells the time of action, and in Standard English there are 12 tenses. The **past tense** is the tense most commonly used to describe events or actions that have already occurred. Therefore, when writing an investigative report, you should use the past tense. For example:

If the Verb Is	*The Past Tense Would Be*
To see	I saw Colin.
To tell	Fava told me.
To say	Ho said he hit Cleworth.
To go	We went to 4th and Orange.
To find	Conk found Egan inside.
To hear	I heard Ito talking.
To smell	We smelled the burning paper.
To write	I wrote the citation.

When quoting someone who is telling you what a suspect said, you should still use the past tense for the speaker. For example:

> Smith told me the suspect said, "Give me your money."
> Brown told me that all Smith said was "Do it."

By writing in the past tense, you will develop consistency and, as a result, create a report that is professional and easy to read.

THE THIRD RULE OF NARRATIVE WRITING—ACTIVE VOICE

The **active voice** is the way of writing that shows who is doing an action, as opposed to the passive voice, which shows who is having something done to them. The writer using the active voice tells who is doing a particular action or thing before telling you who is doing the action or thing. For example:

Active	*Passive*
I wrote the report.	The report was written by me.
I arrested Brown.	Brown was arrested by me.
Smith searched the car.	The car was searched by Smith.
I found the gun in the trunk.	The gun was found in the trunk by me.

One consideration for choosing the active voice over the passive voice is that it is almost always possible to write something in the active voice in fewer words than it would take to write the same thing in the passive voice. A fairly common problem that occurs when investigators use the passive voice is that they forget to include who is doing the action, especially with regard to the chain of custody. It is not uncommon to see sentences such as:

A bag of marijuana was found in the trunk.

Simmons was arrested.

Cape was read his Miranda rights.

The evidence was booked at the central property room.

THE FOURTH RULE OF NARRATIVE WRITING—CHRONOLOGICAL ORDER

One of the most difficult parts of any writing project is getting started, and this is certainly true when it comes to investigative report writing. One popular school of thought says to begin writing at the point of some action being taken, such as what a witness or a victim tells you about the case you are investigating. By doing so, you will avoid having to repeat any information that is contained on the face sheet, such as the date of occurrence or the location. Although this may save some time and paper in the short run, it fails to account for the benefits of starting the report in a way that has long-term and important benefits. One of the keys to successful investigative report writing is to develop consistency, and nowhere is consistency more important from the perspectives of both the reader and the content than at the beginning of the narrative. Here are several benefits to writing a report in **chronological order**.

First, starting a report with the date, time, and how you got involved removes the problem of how to start writing. Second, it clearly establishes the reason an investigation was initiated. Granted there may be incidents where information on the face sheet is repeated in the narrative, but these occurrences will be minimal. Although it will be redundant, the benefits of beginning the report in this manner outweigh the costs of doing so.

A third reason for beginning the narrative of all reports in this manner is that not all investigative reports have face sheets. If one were to follow the suggestion and reasoning that no information should be repeated, there would be two rules to follow—one for reports with face sheets and one for reports without. Having one rule to follow and one way to start the narrative section of a report simplifies the investigator's job, and chances are that the investigator will produce a better work product.

A fourth reason is that the big picture must be considered. In some instances, several investigators will write reports about a single investigation. Homicides or robberies might involve many investigators, all of whom may be trying to gather information about the same case but doing so at different locations and times. At some point, it may become necessary to incorporate all of the individual reports into an overview for a court presentation or a search warrant affidavit. The task of doing either of these is made much easier when the initial sentences of the reports contain all of the needed information to place them in the proper order. Timelines or flowcharts of the investigation are also easily constructed when the date and time are readily accessible. No one likes having to comb through lengthy written passages to find one or two important facts. Starting all reports in this manner eliminates the problem of how to find the beginning of the investigation.

Although those who argue that it is redundant to begin the narrative portion of a report with the date, time, and how you got involved are sometimes correct in their belief, it is still beneficial to start a report this way. The positive aspects of doing so clearly outweigh the negatives. Some examples of the proper way to begin a narrative section are:

> On 12-14-2006 at 0805 hours, I was driving west on 3rd Street when I saw Jones throw a brick through the front window of the house at 4587 3rd Street.
>
> On 7-7-2006 about 1910 hours, I received a radio call to investigate a robbery at the Big Burger Drive-In, 123 Main Street.
>
> On 3-6-2006 at 1245 hours, Sergeant Thompson told me to investigate a reported child endangerment at 6703 Ocean, Apartment 3.
>
> On 12-22-2006 at 1500 hours, I was driving through the south parking lot of the Applewood Shopping Center and saw Brennon run out of Play the Records.

Using specifics to begin the narrative is preferable to the method of using overworked phrases, such as:

> On the above date and time . . .
>
> Witness Wilson arrived home and . . .
>
> Victim states that . . .

It is important to remember that consistency is a key to success and beginning all reports in a consistent manner provides this important feature.

THE FIFTH RULE OF NARRATIVE WRITING—SHORT, CLEAR, CONCISE, AND CONCRETE WORDS

All writers have the option to choose which word to use to describe what they want their readers to understand. As an investigative report writer, you should always strive to get the most out of the words you choose as you write reports. If the ends of the word spectrum are concrete at one end and abstract at the other, the abstract words have no specific meaning and are open to interpretation. **Concrete words** are those that have a clear meaning and little or no misinterpretation of their use.

Abstract words are those that could have multiple meanings depending on the context of a sentence or the reader's viewpoint. For example, the word CONTACTED is widely used in investigative reports and has several potential meanings when used in the sentence:

I contacted Smith.

Does the word "contacted" mean that the investigator touched Smith physically, as in the sense that professional football is a contact sport? Does the word imply that the investigator knows Smith is there in the sense that a radar operator has contact with an airplane as it enters his or her control zone? Does it mean the investigator spoke to Smith?

Concrete words are the opposite of abstract words and are generally better suited for investigative report writing. Words such as "talked," "saw," "found," "searched," and "drove" are examples of concrete words that investigators should use whenever possible. When you use concrete words, you will be able to more clearly describe what you have done or discovered in your investigation. For investigative report writing purposes, writers should always strive to write at the lowest level of abstraction, which means they should use concrete words whenever possible. Other examples of commonly used abstract words and their concrete counterparts are:

abstract: I detected the odor of burning marijuana.
concrete: I smelled burning marijuana.
abstract: We proceeded to the jail.
concrete: We drove to the jail.
abstract: I observed Murphy driving north on Main.
concrete: I saw Murphy driving north on Main.
abstract: Rogers indicated he stole the money.
concrete: Rogers said he stole the money.
abstract: I found the weapon under the couch.
concrete: I found the revolver under the couch.

The five basic rules of narrative writing can be applied to any report writing situation and will allow for the successful completion of any investigative report. They were developed to help investigators achieve consistency in their writing and allow them to complete the writing task as quickly as possible.

When considering all of the things that can be done to improve writing style, it becomes apparent that many areas provide an opportunity for improvement. In addition to the five basic rules of narrative writing, the investigator can do other things to enhance the report and improve its professional quality.

OTHER WRITING CONSIDERATIONS

In addition to the five rules of narrative writing, writers can use five other general guidelines to keep their writing crisp, clean, and to the point. These guidelines include spelling correctly, minimizing the use of abbreviations, properly using names and titles in the narrative, avoiding the use of radio code and jargon, and correctly using direct and indirect quotes.

Spell Correctly

The answer to the question of how to fix an investigator's **spelling** problem is not to merely give him or her a dictionary. The solution lies in doing some basic things and doing them well. In addition to providing the reader a clear description of what the investigator is trying to describe, the use of concrete words will provide another benefit to the writer. Generally, the shorter the word, the easier it will be to spell correctly and the less likely it will be for the writer to make a mistake. The issue of misspelled words in a report is always present, but fortunately, there is an easy way to overcome the problem.

The first step is to use the shortest word possible to say what you mean. The fewer letters the word has, the less chance you will have of making a mistake. The second step is for the investigator to develop a list of words that she or he personally has trouble spelling. The list should be clear, neat, and usable and show the correct spelling of the 100 most problematic words the investigator normally uses in his or her reports. This list can be modified and updated as needed and should always be available. Even though many people would argue that all an investigator needs is a good dictionary to solve a spelling problem, I disagree. Dictionaries usually tell you how to use a word or what the word means. The key to finding out either of these pieces of information is knowing how to spell the word in the first place, to be able to look it up. The third step is to build a bigger vocabulary of short, clear, concise, and concrete words that you can spell.

Minimize the Use of Abbreviations

Another area that report writers need to be aware of is the use of **abbreviations**—or more appropriately, not using abbreviations. Many investigators believe they can save time by using abbreviations because doing so reduces the number and length of words in their report, and accordingly, the report will be shorter.

The end result of such thinking is that the report does end up shorter than had the words been spelled out in full; however, along with a shorter report, there is the fact that many people who read the report will find it confusing. Although it is not realistic to avoid using abbreviations in all situations, they should be avoided in favor of a complete spelling whenever possible. Instead of using an abbreviation, find a complete word that means the same thing. An example of this is the standard law enforcement and investigative word "approximately." Many investigators would not hesitate to use this word in a variety of situations, nor would they hesitate to abbreviate it as "approx." To abbreviate it in this manner requires making seven characters, including the period at the end. If the same investigator used the word "about" in place of "approximately," he or she would use fewer characters and have the same meaning. As we know, abbreviations are appropriate in some situations, and it makes sense to use them. In these cases, use a standard abbreviation that is approved by your agency or company. Use the complete word if there is no commonly used abbreviation. Do not create an abbreviation, but if you must use one, be careful. No one will disagree that a short report takes less time to complete than a long one, but this does not mean that you should sacrifice clear meaning for the sake of brevity.

Use Last Names Without Titles

Another problem for investigative report writers is how to refer to the people they write about. When and how to use **names and titles** can be a dilemma. One of the oldest traditions in investigative report writing is to refer to the people involved in the investigation by titles such as "the victim," "the reporting party," or "witness Fava." Instead of using these titles, use the person's last name. For example, "Brown said," or "I was talking to Kelly," or "I saw Snyder using the shredder." There is nothing wrong with calling people by their names. In fact, it will make the report easier to understand and be more readable. You should also avoid using the titles of Mr., Mrs., and Ms. Just use the person's last name and let it go at that. By eliminating titles before a person's surname, you will save a lot of writing and keep the report looking a lot neater. Remember, if the report is neat and easy to read, it is more useful than one that is cluttered and sloppy.

You can also avoid using the titles of law enforcement officers after the first time the person is mentioned in the report. If a person is listed on the face sheet or some other type of cover page, it is proper to refer to him or her by last name only. There is no need to use the person's title.

There may be instances when you talk to a person who has such minimal, or even no, involvement in a case that he or she should not be listed on a face sheet, or the report you are preparing has no face sheet. The names of these persons should be included in the narrative and as completely identified as possible the first time they are mentioned. From that point on, it is proper to refer to him or her by last name only. For example:

> I spoke to John Brown, 1631 Ninth St., Los Angeles, CA 90085, (213) 555-1212, and he told me he did not know Johnson.

Should it be necessary to refer to this person at a later time in the report, it is appropriate to use the last name only. For example:

> After speaking to Colin, I called Brown at his home and he told me he had not been at the hospital the night of the theft.

Sometimes more than one person with the same last name will be involved in the investigation. In these cases, use both the full first and last names, not the first initial and last name. You are trying to create a document that can be read in the same way a conversation is heard. Not many people will tell you that they saw J. Brown at the market yesterday; they are much more likely to tell you they saw John Brown at the market. Although you should not write the way people speak, you should try to write in a conversational style. Do not use first names only. This is not professional.

When you have a suspect and do not know his or her name until you are well into the investigation, there is always the question of how to refer to the person in the report. A good example of this situation is a driving under the influence arrest. Normally, these cases begin when an officer stops a car being driven erratically and has a conversation with the driver. In this example, the driver, David Thomas, was arrested by Officer Joe Fava for the crime of driving while under the influence of alcohol. After booking Thomas, Fava begins writing a report. Fava would be correct in beginning the report with the date, time, and how he got involved and in so writing, using Thomas's name. For example:

> On 6-13-2006 at 2145 hours, I saw Thomas driving a red Pinto, license 123 ASD, north on Main Street from 5th.

In most cases, going into great detail about how you identified a person is not necessary. If the suspect's identity is crucial to the commission of the crime, a corresponding level of information is needed to establish identity. This, however, is the exception rather than the rule in most cases. Such phrases as "I saw the suspect, later identified as Beer, driving in an erratic manner" are not needed. It is proper to begin by writing, "I saw Beer driving in an erratic manner." Another example of this is the case in which police officer George Najjar gets anonymous information that a person is going to rob the Brand X liquor store sometime after 1:00 PM the following day. Najjar begins a surveillance of the store and, several hours after the investigation begins, sees a person, whose name he later learns is James Dalton enter the store and rob it. The information was received at 7:00 PM on July 3, 2006. Following the rules of narrative writing, the report would begin:

> On 7-3-2006 at 1900 hours, I received an anonymous telephone call in which the caller told me that the Brand X liquor store would be robbed sometime after 1300 hours the next day. On 7-4-2006 about 1100 hours, I began watching the store, and at 1400 hours I saw Dalton walk into the store and. . . .

In this example, the officer did not know Dalton's name when he first saw Dalton go into the store, but it is acceptable to use his name in the report. The rule of thumb is that if you know the person's name before you start writing the report, use it from the beginning. It makes for a more readable report.

Avoid Radio Code and Investigative Jargon

Nearly every occupation or profession has words, expressions, and phrases particular to it that are clear to those in that field but have no meaning to those outside the occupation. The investigative field is no exception. **Jargon** is used daily to give meaning and direction to the activities of investigative personnel, but they do not clearly convey the same message to those outside the field. This unfamiliar ground and the multiple meanings of the words and phrases make their use in investigative reports undesirable and improper. Remember that you are writing for a wide audience and that most people in the audience probably do not understand the vernacular of your specific field. As such, do not use these expressions and phrases.

The use of radio code is another area full of opportunities for confusion and miscommunication. **Radio code** is used to communicate over public airwaves in a shortened way that reduces the airtime needed to transmit a certain message. Radio code is appropriate when a person is talking on the radio or involved in a conversation when others be present should not hear what is being said and there is no other way to safeguard the privacy of the conversation. An example of this would be when an overheard remark would jeopardize an investigation or reveal and compromise what was about to happen in an investigation when timing was crucial to its success. In theory, all investigative agencies that adopt certain radio codes also adopt the same translations. It is important to realize that all investigative agencies do not use the same radio code. In reality, law enforcement agencies within twenty-five miles of each other using the same radio codes many times have different interpretations of those codes. Officers and investigators from these agencies might both hear a familiar code but each would interpret the message differently. If this is the case with professional police officers and investigators, one can only imagine what effect the use of radio code in an investigative report might have on those who are unfamiliar with that specific radio code, or radio code in general, and who read the report. Investigative reports can be confusing enough without the added burden of having to decipher unnecessary radio codes. Remember that your job is to report what happened and what you did so that everyone will understand what is going on.

Use Direct and Indirect Quotes

Quotations are also a concern for the investigative report writer. When to quote and how to write it in the report are substantial issues to resolve. The majority of investigative report writing involves reporting information gained through one of the senses. Few people have the ability to listen to someone and recreate the entire conversation at a later time. More likely, a witness will remember the important things, from his or her perspective, that someone said or told him or her. The witness will tell these important details to the investigator whose job it is to write about them.

The two types of quotes used in investigative report writing are direct and indirect. **Direct quotes** should be used when it is important to know exactly what was said. Many suspects use the same words or phrases when they commit crimes. This then becomes part of their modus operandi, and as such, quoting what was said is important. It is also important to quote what a suspect says when admitting guilt during an interview. There is no better way to describe what a suspect said or did in the commission of a crime than by using his or her own words. Accordingly, there are two times when it is very important to quote directly:

1. When a witness tells you what a suspect said during the commission of a crime.
2. When a suspect admits guilt.

Otherwise, using an **indirect quote** or a paraphrase of what someone says is appropriate. The reason it is not crucial to quote everyone all of the time is that if it becomes necessary for the witness to testify in court or at an administrative hearing about what he or she saw or heard, the witness will be the one to do it. As the investigator, your role will be to act as an assistant to the prosecutor or person presenting the case. Generally, you will not be allowed to testify about what someone told you, so quoting verbatim is not needed. Quoting someone indirectly is giving the essence of what she or he told you. The general idea of what someone said is what you are after, not the exact words. You should remember, however, that if you have the opportunity to quote what a suspect is telling you and you do not, you are giving the advantage to the suspect.

SUMMARY

Writing the narrative portion of a report is often the most challenging part of the investigative report writing process. The task can be made simpler if the investigator uses facts instead of opinions and follows the rules of narrative writing, which guide him or her from the starting point to the conclusion of the report. Writing in the first person and the past tense; using the active voice; keeping things in chronological order starting with the date, time, and how they got involved; and always using short, simple, and concise words is a formula for success in any investigative report. The rules of narrative writing will work in any investigative setting and will be enhanced if the writer also uses correct spelling, avoids abbreviations, correctly and consistently uses names and titles in the narrative, avoids the use of confusing radio code and jargon, and uses direct quotes only when necessary.

REVIEW

1. Use facts, not opinions.
2. Start the narrative of all reports the same way.
3. The five rules of narrative writing are:

   Write in the first person.
   Use the past tense.
   Use the active voice.
   Start with the date, time, and how you got involved.
   Use short, simple, concise, and concrete words.

4. Avoid using abbreviations.
5. Refer to people by their last names.
6. Avoid using titles such as Mr., Mrs., and Ms.
7. Keep radio code and jargon out of the report.
8. Use direct quotes only when needed.

EXERCISES

1. Visit a library and review several publications dealing with trends in writing. Is the continuous, free-flowing, narrative style of writing as described in this chapter discussed in any of your readings? Are there any differences between the rules discussed in your outside readings and those in this chapter? If so, list and discuss them.

2. Check a local newspaper and review the letters to the editor section. Are the writers making their points with facts or opinions?

3. Locate an investigative report that is not written according to the rules of narrative writing. What differences do you see? How can these differences be corrected?

4. What is the difference between the first- and third-person styles of writing? What are the advantages of using the first-person style in investigative reports?

5. Why should the use of jargon be avoided in investigative reports? Give some examples.

6. First-person exercise. Rewrite these sentences in the first-person style. For the purposes of the exercise, you are Officer Fava and your partner is Officer Ho.

   a. U/S officer found the pistol in the street.
   b. Sergeant Najjar gave the evidence to this officer.
   c. Undersigned Officer Fava interviewed Brown.
   d. Officer Fava and Officer Ho searched the car.
   e. U/S officers were assigned to guard the scene.
   f. Cleworth gave the video to Officer Fava.
   g. Officer Donald and Officer Tatum showed this officer where they found the stolen diamond.
   h. It was Investigating Officer Fava's intent to question Jackson as soon as possible.
   i. Officer Fava's and Officer Ho's report is in the watch commander's office.
   j. Officer Ho and I investigated the theft.

7. Past-tense exercise. Rewrite the following examples in the past tense.

   a. I am going to the hospital.
   b. We are finding the evidence.
   c. Booth analyzes the evidence.
   d. Last Tuesday, I was watching the intersection.
   e. After briefing, we proceed to the central jail.

8. Past-tense exercise. Choose the correct word to make the sentence read in the past tense.

   a. At 1130 hours we (eat, ate) lunch.
   b. You (see, saw) the marijuana plants.
   c. We (arrest, arrested) Roberts at 123 Main.
   d. I (book, booked) the knife at central property.
   e. We (drive, drove) to the hospital.

9. Active-voice exercise. Rewrite the following sentences in the active voice.

   a. The call was answered by Rogers.
   b. Smith was read his Miranda rights by me.
   c. The car was searched by Stone.
   d. Ten area cars were wanted by the chief.
   e. The gun was found by Thompson.
   f. The report was written by the lieutenant.

    g. Cleveland was booked by Winters.

    h. The longest speech was given by the mayor.

    i. The window was broken by Watson.

    j. The evidence was examined by Wilson.

10. Chronological order exercise. If not shown, use today's date and military time to establish the proper starting sentence for a narrative report.

    a. It was 3:00 PM when I was dispatched to the Mercy Hospital emergency room.

    b. On Sunday, the 17th of January, approximately 2:00 AM, I found a person, later identified as Sproul, sitting on the steps of the East Branch Library.

    c. Roberts hailed me on March 15, 2006, at about 8:15 AM, while I was driving by the gas station.

    d. Unit 12, which I was driving, was radioed to go to Main and Temple at 6:30 PM to see a man about a theft.

    e. At 1605 hours on April 6, 2006, I was dispatched to the Big Burger Drive-In regarding a disturbance.

    f. On 7-4-2006 at about 2215 hours, we saw Potts throw a lighted flare into a dumpster at the cardboard recycling plant, 1641 West Street.

    g. On 24 March 2006 at 2:05 AM, I heard a traffic collision at Washington and Dysart.

    h. On 12-24-2006 at 2100 hours, I was dispatched to Highway and Noble Drive regarding a theft of Christmas trees in progress.

    i. On Tuesday, February 16, I saw Donaldson driving west on Elm approaching Pacific.

    j. On 5-26-2006 at approximately 1305 hours, Smith assigned me to investigate a theft at the County Credit Union.

11. Short, clear, concise, and concrete word exercise. Suggest a better word to be used in place of each abstract word shown.

    a. proceeded

    b. contacted

    c. detected

    d. advised

    e. indicated

    f. weapon

    g. gun

    h. demonstrated

    i. stated

    j. observed

1. What are the two types of quotes used in investigative reports?

    a. Accurate and exact
    b. Exact and indirect
    c. Gist and accurate
    d. Direct and accurate
    e. Direct and indirect

2. With regard to investigative report writing, what does narrative writing mean?

3. When should abbreviations be used in investigative report writing?

4. List the five rules of narrative writing.

5. How should the narrative section of a report begin?

6. What are three reasons for beginning a narrative in this manner?

7. Describe the first-person style of writing.

8. Why should reports be written in the past tense?

9. Describe the active voice.

10. What is the rule of thumb regarding the use of radio code in the narrative of an investigative report?

5

Describing Persons and Property

In this chapter, you will learn how to categorize the types of people you will encounter when conducting an investigation and completing a report and when to list someone as a suspect. You will also learn how to write a good description of a suspect and of property, both of which can be key in solving a case. Last, you will learn to take care of the evidence in a case by establishing a written record of its custody in an evidence report.

KEY TERMS

Average person test	Replacement cost method
Chain of custody	Reporting partys
Evidence report	Suspect
Fair market value	Victim
Original cost method	Victim appraisal
Others	Witness

Crime fighters on the silver screen, and for that matter the big screen television sets in our living rooms, always seem to make the art of investigation look easy, including the documentation of suspect descriptions and those of the property that is damaged or stolen. Crime fighters in the trenches know it is not that simple.

CATEGORIZING PEOPLE FOR A REPORT

Report face sheets have spaces for a variety of people who have a part in the investigation process. These people include victims, reporting parties, suspects, witnesses, and, in some cases, a large uninvolved group known as others. Who are these people, and how are they defined? When is it appropriate to list someone as a suspect, and when is it not? Who is a witness, and when does a person qualify as an "other"?

Victims

A **victim** is someone who has been hurt or who may have had some of her or his property damaged or stolen. Victims are people who have been the recipient of some wrongful act or deed. They may be persons or entities such as school districts, businesses, or corporations. Although the victim is almost always present when an investigator is looking into a crime, it is not necessary for the victim to be present when the investigator documents the event in a crime report. Neither is it necessary to have the victim present to convict someone of a crime. The victim will almost always identify himself or herself when the investigator arrives and begins the preliminary work of finding out what happened because many times the victim is the person who reported the crime to the investigative agency. In these cases, the victim and the reporting party are one and the same.

Some crimes are categorized based on the characteristics of the victim and of the crime. For example, some burglars only attack residences, whereas others specialize in commercial establishments or businesses. Some burglars operate only at night and commit crimes when people are inside the location they are burglarizing. The three major categories of burglary are residential, commercial, and vehicle. The distinction is determined by the type of target the burglar selects. Persons who commit robberies also often specialize in a certain type. For example, some only attack persons who are on the street, whereas others rob businesses or residences. Therefore, from an investigative standpoint, the categories of robbery are street, residential, and commercial.

A rule of thumb for determining whether the victim of a crime is a person or a business is to determine if the damaged or stolen property belonged to an individual or to a business. This is an area in which department guidelines and the investigator's experience will help define the victim.

It may be possible to have both commercial and personal victims involved in the same crime. Consider the robber who enters a bank and, during the ensuing robbery, takes money belonging to the bank as well as the personal money belonging to the customers inside the bank. If this happened, there would be several victims, including the bank and each person robbed. In all probability, an agency's guidelines determine how the crime is classified. In accordance with these guidelines, the guidelines determine who would investigate it.

Reporting Parties

The person who reports the crime to an investigative agency is called the **reporting party** or person reporting. This person may have no direct knowledge of the crime or know anything about it other than he or she was in a position to summon help. With the increased popularity of cellular phones, it is becoming more and more common to have people drive by something they think is a crime and anonymously report it to a law enforcement agency. The uninvolved person who is walking by a store and hears a person inside yelling for someone to call the police may do so even without knowing if a crime has been committed. The reporting party can also be involved as a victim or as a witness and be within the scope of both of these two categories on the crime report face sheet. When this occurs, it should be noted because one of the purposes of the crime report face sheet is to organize information.

Suspects

The decision to list someone as a **suspect** in a crime report is often determined by agency guidelines that correspond to an internal auditing and investigation procedure. Generally, this auditing and investigative procedure dictates that suspects be divided into two categories—felons and misdemeanants. With regard to felony crime investigations, you should be aware that in many jurisdictions, a felony suspect may be arrested on suspicion that he or she committed a crime. As such, listing a person as a suspect in a felony crime report may be reason enough to arrest him or her if found by a police officer. Therefore, a person should be listed by name on a felony crime report only if there is sufficient probable cause articulated in the report to justify an arrest. If sufficient probable cause does not exist to justify an arrest, the person should not be listed as a named suspect. It would be appropriate in this case to write "SEE NARRATIVE" in the suspect area of the face sheet and put the suspect information in the text of the report. This will allow the information to get to the follow-up investigators while protecting the person's rights. (See Figure 5-1.)

Just as in the case of a felony, when a misdemeanor crime report is being completed, a person should be named as a suspect only when sufficient information exists to justify her or his arrest. In most jurisdictions, a peace officer cannot arrest a person for a misdemeanor that is not committed in his or her presence. Because of this, it would generally not be a problem to list someone as a misdemeanor suspect even if this level of probable cause was absent, but the successful investigative report writer is striving to develop consistent habits and ways of doing things. Developing this consistency makes the task of writing reports much easier. One way in which this report writing simplification will be readily apparent is to have a common method for listing suspect information, regardless of the circumstances.

Witnesses

The test to determine whom a witness might be is an entirely different matter. A **witness** is someone who has useful information about a crime. Ideally, a witness would have this useful information because he or she became aware of it through the use of one or more of the five senses: seeing, hearing, touching, tasting, or smelling something. This, however, is not always the case. A witness may be someone who is aware of evidence of a crime because he or she saw it occur or because someone who witnessed the crime told the witness where evidence might be located. When a person is found who has useful information about a crime being investigated he or she should be listed as a witness. Useful information may be defined as anything that helps solve a crime, points to or eliminates a particular person as being involved, or identifies property or evidence in a given case, or anything else that might be important to the investigation.

Others

Not all people an investigator talks to during an investigation can be categorized in one of the areas already discussed. There will be many people who have no information, who did not see or hear anything at all relative to the matter being investigated, and whose only reason for being contacted by the investigator is that they were in the same area as the investigator at a time he or she wanted to talk to them. Persons in this category are known as **others** in report writing vernacular. Not only is there a good reason for including them in the report, but there is also a proper way to do so. An example involving an "other" is:

A car parked at a shopping mall is vandalized by someone who throws a brick onto the hood, denting the metal and scratching the paint. The police are called, and while the officer talks to Robert Smith, the owner of the vandalized car, Brenda Wilson, who owns the car parked next to Smith's, arrives at her car. The officer asks Wilson if she saw anything unusual when she arrived at the mall. Wilson tells the officer that she arrived two hours before Smith did and was shopping inside the mall when Smith arrived.

EL SEGUNDO POLICE DEPARTMENT	SUSPECT REPORT	PAGE_____ OF _____
		CASE NO

CRIME 1

CODE SECTION		CRIME		CLASSIFICATION				REFER OTHER REPORTS	
LOCATION (Be Specific)				RD.	DATE	TIME	SUPPL. ☐	INCIDENT NO.	

SUSP. VEH. 2

LICENSE #		STATE	VEH. YR	MAKE	MODEL	BODY STYLE ☐ 0 UNK ☐ 2 4-DR ☐ 4 P/U ☐ 6 VAN ☐ 8 RV ☐ 10 OTHER ☐ 1 2-DR ☐ 3 CONV ☐ 5 TRUCK ☐ 7 S/W ☐ 9 M/C
COLOR/COLOR		OTHER CHARACTERISTICS (i.e. T/C Damage, Unique Marks or Paint, etc.)			DISPOSITION OF VEH.	
REGISTERED OWNER						

SUSPECT 3

SUSP. #	NAME (First, Last, Middle)	SEX ☐ 1. M ☐ 2. F	RACE ☐ 0 UNK ☐ 2 HISP ☐ 4 IND ☐ 6 JAP ☐ 8 OTH_____ ☐ 1 WHT ☐ 3 BLK ☐ 5 CHI ☐ 7 FIL ☐ 9 P.ISL.

AKA	D.O.B.	AGE	HT.	WT.	BUILD ☐ 0 UNK ☐ 2 MED ☐ 4 MUSCLR ☐ 1 THIN ☐ 3 HEAVY

HAIR ☐ 0 UNK ☐ 2 BLK ☐ 4 RED ☐ 6 S/P ☐ 8 OTHER _____   ☐ 1 BRN ☐ 3 BLN ☐ 5 GRAY ☐ 7 WHT _____

EYES ☐ 0 UNK ☐ 2 BLK ☐ 4 GRN ☐ 6 GRAY   ☐ 1 BRN ☐ 3 BLU ☐ 5 HAZEL ☐ 7 OTHER _____ D.L. #

RES. ADDRESS	RD	ZIP CODE	RES. PHONE # ( )	S.S #
BUS. ADDRESS	RD	ZIP CODE	BUS. PHONE # ( )	OCCUPATION

CLOTHING	ARRESTED ☐ 1 YES ☐ 2 NO	STATUS ☐ 1 DRIVER ☐ 3 PED. ☐ 2 PASS	GANG AFFILIATION: HOW KNOWN:	☐ 1 KNOWN ☐ 2 SUSPECTED

AMT. OF HAIR 4	HAIR STYLE 8	COMPLEXION 10	TATTOOS/SCARS 13	DISTING. MARKS 14	WEAPON(S) 17
☐ 0 UNKNOWN Q21	☐ 0 UNKNOWN Q25	☐ 0 UNKNOWN Q27	☐ 0 UNKNOWN	☐ 0 NONE Q30	☐ 0 UNKNOWN ☐ 0 NONE Q33
☐ 1 THICK	☐ 1 LONG	☐ 1 CLEAR	☐ 1 FACE	_____	☐ 1 CLUB _____
☐ 2 THIN	☐ 2 SHORT	☐ 2 ACNE	☐ 2 TEETH	_____	☐ 2 HAND GUN _____
☐ 3 RECEDING	☐ 3 COLLAR	☐ 3 POCKED	☐ 3 NECK	_____	☐ 3 OTHER UNK GUN _____
☐ 4 BALD	☐ 4 MILITARY	☐ 4 FRECKLED	☐ 4 R/ARM	_____	☐ 4 RIFLE _____
☐ 5 OTHER_____	☐ 5 CREW CUT	☐ 5 WEATHERED	☐ 5 L/ARM	_____	☐ 5 SHOT GUN _____
TYPE OF HAIR 5	☐ 6 RIGHT PART	☐ 6 ALBINO	☐ 6 R/HAND	_____	☐ 6 TOY GUN _____
☐ 0 UNKNOWN Q22	☐ 7 LEFT PART	☐ 7 OTHER_____	☐ 7 L/HAND	_____	☐ 7 SIMULATED _____
☐ 1 STRAIGHT	☐ 8 CENTER PART	**GLASSES 11**	☐ 8 R/LEG	_____	☐ 8 POCKET KNIFE _____
☐ 2 CURLY	☐ 9 STRAIGHT BACK	☐ 0 UNKNOWN Q28	☐ 9 L/LEG	_____	☐ 9 BUTCHER KNIFE _____
☐ 3 WAVY	☐ 10 PONY TAIL	☐ 0 NONE	☐ 10 R/SHOULDER	_____	☐ 10 OTH. CUT/STAB INST _____
☐ 4 FINE	☐ 11 AFRO/NATURAL	☐ 1 YES (No Descrip.)	☐ 11 L/SHOULDER	_____	☐ 11 HANDS/FEET _____
☐ 5 COARSE	☐ 12 PROCESSED	☐ 2 REG GLASSES	☐ 12 FRONT TORSO	_____	☐ 12 BODILY FORCE _____
☐ 6 WIRY	☐ 13 TEASED	☐ 3 SUN GLASSES	☐ 13 BACK TORSO	_____	☐ 13 STRANGULATION _____
☐ 7 WIG	☐ 14 OTHER_____	☐ 4 WIRE FRAME	☐ 14 OTHER	_____	☐ 14 TIRE IRON _____
☐ 8 OTHER_____	**FACIAL HAIR 9**	☐ 5 PLASTIC FRAME	_____	_____	☐ 15 OTHER _____
HAIR CONDITION 6	☐ 0 UNKNOWN Q26	☐ Color_____			**WEAPON FEATURE 18**
☐ 0 UNKNOWN Q23	☐ 0 N/A	☐ 6 OTHER_____	**UNIQUE CLTHNG 15**	**WEAPON IN 16**	☐ 0 UNKNOWN ☐ 0 NONE Q34
☐ 1 CLEAN	☐ 1 CLN SHAVEN	**VOICE 12**	☐ 0 UNK ☐ 0 NONE	☐ 0 UNKNOWN Q32	☐ 1 ALTERED STOCK _____
☐ 2 DIRTY	☐ 2 MOUSTACHE	☐ 0 UNKNOWN Q29	☐ 1 CAP/HAT Q31	☐ 0 N/A	☐ 2 SAWED OFF _____
☐ 3 GREASY	☐ 3 FULL BEARD	☐ 0 N/A		☐ 1 BAG/BRIEFCASE	☐ 3 AUTOMATIC _____
☐ 4 MATTED	☐ 4 GOATEE	☐ 1 LISP	☐ 2 GLOVES	☐ 2 NEWSPAPER	☐ 4 BOLT ACTION _____
☐ 5 ODOR	☐ 5 FUMANCHU	☐ 2 SLURRED	_____	☐ 3 POCKET	☐ 5 PUMP _____
☐ 6 OTHER_____	☐ 6 LOWER LIP	☐ 3 STUTTER	☐ 3 SKI MASK	☐ 4 SHOULDER	☐ 6 REVOLVER _____
R/L HANDED 7	☐ 7 SIDE BURNS	☐ 4 ACCENT	_____	HOLSTER	☐ 7 BLUE STEEL _____
☐ 0 UNKNOWN Q24	☐ 8 FUZZ	Describe_____	☐ 4 STOCKING MASK	☐ 5 WAISTBAND	☐ 8 CHROME/NICKEL _____
☐ 1 RIGHT	☐ 9 UNSHAVEN	_____	_____	☐ 6 OTHER_____	☐ 9 DOUBLE BARREL _____
☐ 2 LEFT	☐ 10 OTHER_____	☐ 5 OTHER_____	☐ 5 OTHER_____		☐ 10 SINGLE BARREL _____
					☐ 11 OTHER _____

REPORTING OFFICER	ID#	DATE	REVIEWED BY	ID#	DATE

COPIES: ☐ CHIEF ☐ CII ☐ PATROL ☐ DB ☐ OTHER AGENCY	ROUTED BY	ENTERED BY
TO: ☐ ☐ CAU ☐ ABC (2 copies) ☐ DA _____		

ESPD Form #200 (Rev 5/97)

Figure 5-1 Suspect Report. (Courtesy of the El Segundo Police Department)

CASE NO.

PAGE ______

SUSPECT 3

SUSP. #	NAME (First, Last, Middle)	SEX ☐ 1 M ☐ 2 F	RACE ☐ 0 UNK ☐ 1 WHT ☐ 2 HISP ☐ 3 BLK ☐ 4 IND ☐ 5 CHI ☐ 6 JAP ☐ 7 FIL ☐ 8 OTH______ ☐ 9 P.ISL.			
AKA		D.O.B.	AGE	HT.	WT.	BUILD ☐ 0 UNK ☐ 1 THIN ☐ 2 MED ☐ 3 HEAVY ☐ 4 MUSCLR
HAIR ☐ 0 UNK ☐ 1 BRN ☐ 2 BLK ☐ 3 BLN ☐ 4 RED ☐ 5 GRAY ☐ 6 S/P ☐ 7 WHT ☐ 8 OTHER ______		EYES ☐ 0 UNK ☐ 1 BRN ☐ 2 BLK ☐ 3 BLU ☐ 4 GRN ☐ 5 HAZEL ☐ 6 GRAY ☐ 7 OTHER ______				D.L. #
RES. ADDRESS		RD	ZIP CODE	RES. PHONE # ( )		S.S #
BUS. ADDRESS (School)		RD	ZIP CODE	BUS. PHONE # ( )		OCCUPATION
CLOTHING	ARRESTED ☐ 1 YES ☐ 2 NO	STATUS ☐ 1 DRIVER ☐ 2 PASS ☐ 3 PED.	GANG AFFILIATION: HOW KNOWN:			☐ 1 KNOWN ☐ 2 SUSPECTED

AMT. OF HAIR 4
☐ 0 UNKNOWN
☐ 1 THICK
☐ 2 THIN
☐ 3 RECEDING
☐ 4 BALD
☐ 5 OTHER______

TYPE OF HAIR 5
☐ 0 UNKNOWN
☐ 1 STRAIGHT
☐ 2 CURLY
☐ 3 WAVY
☐ 4 FINE
☐ 5 COARSE
☐ 6 WIRY
☐ 7 WIG
☐ 8 OTHER______

HAIR CONDITION 6
☐ 0 UNKNOWN
☐ 1 CLEAN
☐ 2 DIRTY
☐ 3 GREASY
☐ 4 MATTED
☐ 5 ODOR
☐ 6 OTHER______

R/L HANDED 7
☐ 0 UNKNOWN
☐ 1 RIGHT
☐ 2 LEFT

HAIR STYLE 8
☐ 0 UNKNOWN
☐ 1 LONG
☐ 2 SHORT
☐ 3 COLLAR
☐ 4 MILITARY
☐ 5 CREW CUT
☐ 6 RIGHT PART
☐ 7 LEFT PART
☐ 8 CENTER PART
☐ 9 STRAIGHT BACK
☐ 10 PONY TAIL
☐ 11 AFRO/NATURAL
☐ 12 PROCESSED
☐ 13 TEASED
☐ 14 OTHER______

FACIAL HAIR 9
☐ 0 UNKNOWN
☐ 0 N/A
☐ 1 CLN SHAVEN
☐ 2 MOUSTACHE
☐ 3 FULL BEARD
☐ 4 GOATEE
☐ 5 FUMANCHU
☐ 6 LOWER LIP
☐ 7 SIDE BURNS
☐ 8 FUZZ
☐ 9 UNSHAVEN
☐ 10 OTHER______

COMPLEXION 10
☐ 0 UNKNOWN
☐ 1 CLEAR
☐ 2 ACNE
☐ 3 POCKED
☐ 4 FRECKLED
☐ 5 WEATHERED
☐ 6 ALBINO
☐ 7 OTHER______

GLASSES 11
☐ 0 UNKNOWN
☐ 0 NONE
☐ 1 YES (No Descrip.)
☐ 2 REG GLASSES
☐ 3 SUN GLASSES
☐ 4 WIRE FRAME
☐ 5 PLASTIC FRAME
☐ Color______
☐ 6 OTHER______

VOICE 12
☐ 0 UNKNOWN
☐ 0 N/A
☐ 1 LISP
☐ 2 SLURRED
☐ 3 STUTTER
☐ 4 ACCENT
Describe______
☐ 5 OTHER______

TATTOOS/SCARS 13 | **DISTING. MARKS 14**
☐ 0 UNKNOWN ☐ 0 NONE
☐ 1 FACE ______
☐ 2 TEETH ______
☐ 3 NECK ______
☐ 4 R/ARM ______
☐ 5 L/ARM ______
☐ 6 R/HAND ______
☐ 7 L/HAND ______
☐ 8 R/LEG ______
☐ 9 L/LEG ______
☐ 10 R/SHOULDER ______
☐ 11 L/SHOULDER ______
☐ 12 FRONT TORSO ______
☐ 13 BACK TORSO ______
☐ 14 OTHER ______

UNIQUE CLTHNG 15
☐ 0 UNK ☐ 0 NONE
☐ 1 CAP/HAT ______
☐ 2 GLOVES ______
☐ 3 SKI MASK ______
☐ 4 STOCKING MASK ______
☐ 5 OTHER______

WEAPON IN 16
☐ 0 UNKNOWN
☐ 0 N/A
☐ 1 BAG/BRIEFCASE
☐ 2 NEWSPAPER
☐ 3 POCKET
☐ 4 SHOULDER HOLSTER
☐ 5 WAISTBAND
☐ 6 OTHER______

WEAPON(S) 17
☐ 0 UNKNOWN ☐ 0 NONE
☐ 1 CLUB ______
☐ 2 HAND GUN ______
☐ 3 OTHER UNK GUN ______
☐ 4 RIFLE ______
☐ 5 SHOT GUN ______
☐ 6 TOY GUN ______
☐ 7 SIMULATED ______
☐ 8 POCKET KNIFE ______
☐ 9 BUTCHER KNIFE ______
☐ 10 OTH. CUT/STAB INST ______
☐ 11 HANDS/FEET ______
☐ 12 BODILY FORCE ______
☐ 13 STRANGULATION ______
☐ 14 TIRE IRON ______
☐ 15 OTHER ______

WEAPON FEATURE 18
☐ 0 UNKNOWN ☐ 0 NONE
☐ 1 ALTERED STOCK ______
☐ 2 SAWED OFF ______
☐ 3 AUTOMATIC ______
☐ 4 BOLT ACTION ______
☐ 5 PUMP ______
☐ 6 REVOLVER ______
☐ 7 BLUE STEEL ______
☐ 8 CHROME/NICKEL ______
☐ 9 DOUBLE BARREL ______
☐ 10 SINGLE BARREL ______
☐ 11 OTHER ______

ESPD Form #200 (Rev 5/97)

Figure 5-1 *(Continued)*

The question is this: What does the officer do with the information from Wilson? Some officers would not include Wilson's information in the report, but others would consider Wilson a witness. The primary question is this: Does Wilson have any useful information about the case? The answer clearly is no, she does not. The next logical question is this: How should Wilson's information be handled in the report? The correct method of handling it is to include it in the narrative in the chronological order in which it was received. It might read:

> While I was talking to Smith, Brenda Wilson, 123 Maine, Pasadena, CA, 90876, (818) 555-1212, arrived and told me . . . .

It is important to include this information in the report for the following reasons:

1. It is appropriate to document all of the investigation you do. Your supervisors might review your reports to see what type of investigations you do as well as evaluate their quality, thoroughness, and the manner in which you proceed.
2. Even if you merely talk to someone, you should put it in the report. Even though the person tells you he or she did not see or hear anything related to the inquiry, it may turn out that the person was somehow involved.
3. It is a way to document what you did, and this adds to your credibility if you are called to testify in court.
4. The person you are talking to who claims to know nothing about the crime may be lying about her or his involvement, although you may be unaware of it at the time. Once the report is forwarded to the investigation division for follow-up, the name of the uninvolved person may mean something to the other investigators who read or hear about the case.

It is important to keep the big picture in mind when you are investigating what seems like a small matter because more than once a person who seems uninvolved today could become a suspect tomorrow.

Describing people in a report is a simple matter if the witness can provide a good description. If, however, no one saw the suspect or saw him or her for only a split second, getting a good description may be more difficult and presents a challenge to you, the investigator.

ASSESSING WEIGHTS AND MEASUREMENTS

To start, you must have a basic ability to assess weights and measurements. For example, you should know how tall you are, as well as the distance from the ground to the tip of your nose, to the top of your shoulders, and to your waist. You should know your weight and the approximate weights of people, both male and female, of various heights with average builds. With this knowledge you can practice guessing the height and weight of people you know until you become proficient. Perfecting this skill will serve you well as you interview witnesses and ask them about people they saw committing crimes you are investigating.

INTERVIEWING FOR SUSPECT DESCRIPTIONS

As you interview people to get suspect descriptions, it may not be realistic to expect a complete description on the first try. What is more likely to occur is that they will tell you the things they remember. The information they will remember will include the things that stood out about the person, in the order of importance to the witness. The order in which they describe the suspect to you will make perfect sense to them, but it will probably not be in the order you need to complete your report. Good investigators have a plan in place that allows them to get the information they need in a timely fashion and allows the witness to report all the information he or she knows.

Now is a good time to remember the basic rule of note taking, which is listen first, then write. Ask the witness to tell you what the suspect looked like and as he or she tells you, just listen. Once finished, ask him or her to answer specific questions based on the order of information you need for your report. A suggested order or formula for recording this information is:

Sex, race, age, height, weight, hair style and color, eye color, clothing description starting at the top with outside garments and working down, and anything else that is of importance such as tattoos, missing limbs or teeth, accents, unusual gait, and so on.

People will not always recall all the information about a suspect even when they had an unobstructed view of the person for several minutes. When you are able to get a full and complete description, it is a simple matter to write it in the report. When several parts of the description are left unfilled, it is acceptable to continue the description with the next piece of known information. For example:

Male, white, about 35, 6' 3", 197 pounds, short brown hair, wearing a white short sleeve sweatshirt, gray shorts, white slip-on shoes.

As you can see, a great deal of information is missing from this description as compared to the ideal formula, yet there is enough information available to begin a search. A description is used not only to help find the suspect but also to eliminate those who are not involved from the suspect pool.

There will be times when a witness provides information that is too lengthy to fit into the boxes on the crime report face sheet.

In these cases, write "SEE NARRATIVE" in the suspect information boxes and then write the suspect's description on the first page of the narrative after labeling the area "SUSPECT INFORMATION." This will give you all the room you need to write a thorough suspect description. Keep in mind that the spaces on the face sheet are designed to guide and help you complete the report, not limit what you write. When it comes to describing suspects, you can never have too much good information.

If you have multiple suspects, start by helping the witness focus on one suspect and get as complete a description as possible before repeating the same procedure for all additional suspects. If you have no names for the suspects, it is acceptable and appropriate to give them numbers such as suspect #1 and suspect #2.

It is rare that several witnesses see a suspect and describe him or her in the same terms. In those cases in which witnesses give different descriptions, you should list the suspect by witness description so that it is clear which witness said what about the suspect. For example:

> Suspect #1 as described by Cleworth:
> male, white, 35 years, 6' 5", 230 pounds, black hair . . .
> Suspect #1 as described by Najjar:
> male, Hispanic, 30 years, 6' 3", 215 pounds, dark hair . . .
> Suspect #1 as described by Ito:
> male, white, mid-30s, 6' 4", 220 pounds, black hair . . .

In this case, the witnesses saw a person they believe to be the suspect, but they all remember him differently. It would be unacceptable to combine the information into one hybrid description for the sake of speed and brevity. This would not help solve the crime and quite possibly would hinder any prosecution if the witnesses became confused about what they saw.

DESCRIBING PROPERTY IN A REPORT

How to record thorough and accurate property descriptions poses two main problems for the investigative report writer. First, the many different brands, sizes, colors, and models of products available create an almost infinite number of things to describe. Second, recording an accurate value for this infinite number of items is difficult because the value can be influenced by appreciation, depreciation, damage, or collectibility. (See Figure 5-2.)

THE AVERAGE PERSON TEST

Describing property is probably the easier of the two tasks addressed in this chapter. There is a simple way to get a quality description of any piece of property, whether it is something that has been reported as stolen by a victim, booked as a piece of

Figure 5-2   Property Report. (Courtesy of the La Habra Police Department)

evidence, turned in as found property, or suspected of being contraband but its status cannot be determined on the spot.

This simple method is known as the **average person test** and is easy for the investigator to apply. Whenever you are describing any piece of property or evidence, do so with the objective of writing a description that is so complete and so thorough that the average person could look at five or six similar items and, based on the description you wrote, pick out the item you have described. If the average person can do this, then the description is good enough. If the average person cannot, the description is not good enough.

PHOTOS AND SKETCHES

One way you can help develop your description to meet the average person test of completeness is to include a photograph or sketch of the missing property. Many people will have photos showing their jewelry and possessions or perhaps a video recording of their valuable items for use in insurance claims. With digital technology available, reproducing still photography and video images is very doable. In these cases, a picture may not be worth a thousand words, but it might be the difference between a piece of recovered property being returned to its rightful owner and its being sold at a public auction as unclaimed. If photos or video is not available, the owner of a piece of jewelry or other valuable may be able to accurately draw a sketch of the item.

DETERMINING PROPERTY VALUE

Assigning a value to stolen property is another area of concern for many investigators because there are several ways to do so. You can spend hours haggling with victims about the value of their five-year-old compact disc player or the intrinsic value of the coin collection they have had since they were children, but it is necessary to have a value for property because the value of the loss is important in determining the corpus, as in the case of the difference between misdemeanor or grand theft. The amount of the loss in theft cases is also used in determining the level of assignment for investigators. Limited resources may prevent some agencies from working theft cases with a minimal loss as diligently as they would a case with a large one.

Fortunately, there are many ways of arriving at value estimates in investigative reports. Some of the most common methods are:

1. **Original cost.** Using this method requires the victim or owner of the property to remember what he or she paid for the item. This amount is used in the report without regard for subsequent damage that might have lessened its value or to appreciation that might have increased its value. This method

sometimes requires the owner to present a receipt showing the purchase price. (See Figure 5-3.)

2. **Fair market value.** This method requires the investigator to use his or her judgment in determining the value of an item by considering what it was worth when it was new, what the current demand for the item is, and what it may be worth now. Interest by the public or of some collectors' group might increase its value. The rarity of the item might also be a factor. When using this method, the investigator makes a judgment based on his or her best informed guess as to the value and includes this in the report.

3. **Victim appraisal.** This method requires the victim to provide the value of the missing property by telling the investigator how much it is worth. The victim may use, among other criteria, what she or he perceives as the replacement cost, historical value, or estimated replacement cost, but the bottom line is that the value is going to be what the victim says. This is a less than scientific way of determining the value of something, but investigators waste a lot of valuable time determining the value of property when it really is not that important to the investigating agency. Unless the value is overreported by thousands of dollars, the net effect on the investigating agency is minimal. The agency is still going to investigate the crime and perform the administrative duties required by law. If the real concern is to report an accurate amount to prevent insurance fraud, it is necessary to consider that insurance companies utilize the services of agents, private investigators, and estimators to verify the value of property taken before paying a claim. These claims adjusters, not the investigator, have as one of their primary responsibilities the duty of determining the value of a loss.

4. **Replacement cost.** Figuring a replacement cost can be tricky. Not only does the investigator need to know where an item can be purchased, but also if there is any kind of rebate plan, sale price, or discount available to the victim.

There are advantages and disadvantages to each of these methods of property value determination. In choosing the method that will be used, it is important to consider agency requirements. If a particular method is used by the agency, the investigator should stay with that method for the sake of consistency. If the investigator has the option of selecting a method, the victim appraisal method is recommended because it quickly establishes a value. If the victim overestimates the value of the loss, the insurance company will correct it at the time the settlement is made and your report can be adjusted accordingly.

Determining the value of property should not be the most involved or important thing an investigator does during a preliminary investigation. It should take a minimal amount of time

EL SEGUNDO POLICE DEPARTMENT
PROPERTY REPORT

DATE AND TIME OF REPORT

DR#

- ☐ ARRESTEE
- ☐ SUSPECT
- ☐ VICTIM
- ☐ WITNESS

- ☐ EVIDENCE
- ☐ FOUND PROPERTY
- ☐ SAFEKEEPING
- ☐ PRISONER PROPERTY

- ☐ DESTRUCTION
- ☐ UNDER OBSERVATION
- ☐ HOLD: CONTACT OFFICER _____
- ☐ OTHER _____

NAME (Property Booked To)

NAME (Secondary Persons)

☐ MISDEMEANOR
☐ FELONY

ADDRESS

ADDRESS

CHARGES

1.

CITY   ZIP

CITY   ZIP

2.

PHONE

PHONE

RES.:   BUS.:

RES.:   BUS.

3.

B I K E	MAKE	MODEL	☐ BOY'S ☐ GIRL'S SPEED 1☐ 3☐ 5☐ 10☐ 15☐ OTHER☐	FRAME COLOR	WHEEL SIZE
	SERIAL NO.	LICENSE NO. CITY	OTHER DESCRIPTIONS		

FOUND PROPERTY STATEMENT: DO YOU WISH TO CLAIM THIS PROPERTY IF THE LEGAL OWNER CANNOT BE LOCATED? ☐ YES ☐ NO
PROPERTY CAN BE CLAIMED AFTER **90 DAYS** DATE_____ FINDER SIGNATURE_____

PRISONER PROPERTY STATEMENT: ITEMS NOT ABLE TO GO TO COURT WILL BE HELD BY THIS DEPARTMENT FOR A PERIOD OF **ONE YEAR.** IF AFTER 1 YEAR YOU HAVE NOT MADE ARRANGEMENTS WITH THE PROPERTY & EVIDENCE OFFICER TO CLAIM YOUR PROPERTY WE WILL DESTROY ALL PRISONER PROPERTY AFTER THAT DATE.
PRISONER SIGNATURE _____ DATE_____

LIST ALL ARTICLES STARTING WITH ITEM NO. 1: DESCRIBE PROPERTY AS FOLLOWS.

ITEM NO.	QTY.	ARTICLE, DESCRIPTION, BRAND NAME, MODEL, ETC.	SERIAL NO.	VALUE OR WEIGHT

CONTINUATION FORM USED FOR ADDITIONAL ITEMS ☐

COMPLAINT FORM NARRATION:

REPORTING OFFICER (Name and Serial No.) | TEMPORARY LOCATION OF PROPERTY | APPROVING SUPERVISOR | DATE

PROPERTY USE ONLY
PROPERTY ROOM LOCATION | BY: | DATE | FINAL DISPOSITION
PROPERTY RELEASED TO: | ADDRESS: | DATE

WHITE: WITH EVIDENCE   **YELLOW:** TO RECORDS   **PINK:** TO FINDER/PRISONER PROPERTY

Figure 5-3 Property Report. (Courtesy of the El Segundo Police Department)

CONTINUATION REPORT

ITEM NO.	QTY.	ARTICLE DESCRIPTION, BRAND NAME, MODEL ETC.	SERIAL NO.	VALUE

DATE · OFFICER · DR#

REPORTING OFFICER (Name and Serial No.)	TEMPORARY LOCATION OF PROPERTY	APPROVING SUPERVISOR	DATE
PROPERTY USE ONLY PROPERTY ROOM LOCATION	BY:	DATE	FINAL DISPOSITION
FCN NO.	BY:		DATE

Figure 5-3 *(Continued)*

and allow him or her to move forward in trying to locate witnesses and physical evidence.

WRITING EVIDENCE REPORTS

Describing property in crime reports is only one application in which you must be accurate and specific in your writing. Another report writing area where excellence in writing is the only acceptable standard is when dealing with evidence in the aptly titled "EVIDENCE REPORT." An **evidence report** has two basic components. The first is a description of the evidence, and the second is the chain of custody.

Evidence can take many forms and come into your care and custody in a variety of ways. If something tends to prove a fact or gives you a basis for believing something, it is likely evidence. How you identify and collect evidence in your specific cases is the subject of another class and another skill set. For the purpose of this book, let's assume that your skills are good and you are using the proper techniques. Now your problem is introducing the evidence into your report and accounting for it as the case continues.

There are many ways to account for evidence, but the two most popular are to refer to the evidence in the narrative of the investigative report in the chronological order it was received, using great detail and accountability. The other is to refer to the evidence in the narrative in a generic sense and then in greater detail in a separate report called the **Evidence Report**. The second method is recommended because it gives greater freedom to the writer in describing the evidence, it creates better organization in the report, and it makes the report easier to read and understand.

DESCRIBING EVIDENCE

Just as with property descriptions, there is no room for error when describing evidence. The description you use in the evidence report must also meet and exceed the average person test. The level of additional detail is needed because you may be asked to identify a specific piece of evidence from the witness stand while someone holds or displays the item several feet from you. This might seem like a big problem, but it is easy to solve if proper marking and packaging techniques are used.

The marking system you select and use should not destroy the evidence or diminish its value. It just needs to be something you can identify as your mark at a later date. The packaging and tagging process should be something that is manageable with as little effort as possible and still give a solid level of protection to the evidence. Packaging materials are commonly found in the evidence preparation rooms of police stations. (See Figure 5-4.)

El Fuego Police Department

Evidence Report   **99-076587**

        On 4-27-99 at 1545 hours a search warrant was served at
623 Fastwater Lane. I found the following items during the
subsequent search.

1.   (1) Walther PPK, .380 Caliber handgun, blue steel finish
      with brown plastic grips. Serial number WP164128.
      Found under couch in the living room.

2. $647.00 U.S. currency consisting of (5) $100 bills, (2)
      $50 bills, (4) 10 bills, and (7) $1.00 bills. Found
      under bed in master bedroom.

3. $1000 U.S. currency consisting of (10) $100 bills.
      Found in freezer wrapped in foil.

4.   (1) clear plastic baggy containing an unknown type
      white powder. 463 grams gross weight.

5.   (1) clear plastic baggy containing marijuana. 1000
      grams gross weight.

I kept all the items with me and marked, packaged, tagged,
and booked them at the El Fuego Police Department.

  M. Colin   #167

Figure 5-4   Sample Evidence Report.

EVIDENCE REPORT FORMATS

Although there are many different crime report face sheets, there are basically two different report formats used for evidence reports. The first format uses a printed form that serves much like a crime report face sheet. It organizes the information and prompts the writer with headings. This is in essence a fill-in-the-blanks report. Although there are differences in the way the form looks from one agency to another, the information needed to complete it is basically the same.

The second format uses a free-flowing narrative on a blank piece of paper, which allows the writer to include as much detail and information about the evidence as is needed. Some agencies have strict guidelines about which items of evidence to list first, second, third, and so on. There are a number of agencies that do not have a protocol for evidence reports. In these cases, a workable solution to the problem is to divide the evidence into general categories and then list them in the following order:

GUNS

MONEY

DRUGS

ITEMS WITH SERIAL NUMBERS

ITEMS WITHOUT SERIAL NUMBERS

These headings would not appear on the evidence report but would act as a guide for you in setting up the report. The items on the evidence report should be consecutively numbered, beginning with the first item seized. This system will allow additional evidence to be added if the case is ongoing. The particulars of where and when the evidence was seized would be shown for each item.

No matter which format of evidence report you use, you must account for all the evidence you seize; in other words, you must establish a chain of custody. (See Figure 5-5.)

ESTABLISHING THE CHAIN OF CUSTODY

The **chain of custody** is the term that describes the handling and care of evidence. When evidence is seized, it must be identified, described, and accounted for from the time it was seized until it is entered into evidence at trial.

Typically, the process or path that evidence takes begins when the investigator finds an item that has real or potential evidentiary value. The investigator collects the item, processes it if necessary, and then marks it for later identification. Once marked, the item is packaged, tagged, and placed in a secure evidence storage facility. Once it has been booked in the evidence facility, the responsibility to care for the evidence rests with the person who manages the facility or laboratory where the investigator puts the evidence. (See Figure 5-6.)

PROPERTY OF:		
DATE:		DR#
DENOM.	NUMBER	AMOUNT
100's		$
50's		
20's		
10's		
5's		
1's		
OTHER		
CHANGE		
TOTAL		$
OFFICER'S SIGNATURE		
VERIFYING OFFICER		
SUBJECT'S SIGNATURE		

PE-028

Figure 5-5 Envelope Used to Book Money. (Courtesy of the El Segundo Police Department)

Establishing the chain of custody is also part of your responsibility as the report writer, and it is properly done in the evidence report. The report numbers each piece of evidence separately beginning at 1, and includes a description of the item, where it was found, who found it, and where it was booked. In general, this establishes the chain of custody and might look like this:

EVIDENCE

Case No. _____ **Inventory #** _____

Type of offense _____

Description of evidence _____

_____

Suspect _____

Victim _____

Date and time of recovery _____

_____

Location of recovery _____

_____

Recovered by _____

CHAIN OF POSSESSION

Received from _____

  By _____

  Date _____ **Time** _____ **AM PM**

Received from _____

  By _____

  Date _____ **Time** _____ **AM PM**

Received from _____

  By _____

  Date _____ **Time** _____ **AM PM**

LYNN PEAVEY COMPANY   800-255-6499

Figure 5-6 Sample Evidence Tag. (Courtesy of the El Segundo Police Department)

On 10-13-2006 at 1130 hours, I found the following item at 640 Peach Street and marked, packaged, tagged, and booked it at the Department of Justice Lab in Sacramento.

1. (1) One-gallon, red, plastic gasoline can, empty, found in the garage.

This example can be expanded to any number of items found at the same scene and allows the writer to keep the chain of custody clear and well established. This also shows the benefit of writing in the active voice.

You may be involved in cases in which several investigators search for and find evidence and then give it to another investigator who is responsible for the report. In this case, the items would be numbered and described just as in the first example, with the addition of who found it. The chain of custody would look like this:

On 12-25-2006 at 1845 hours, conducted a search at 123 Main Street and the following evidence was found:

1. (1) one-gallon, red, plastic gasoline can, empty, found in the garage by Ho.
2. (1) partially burned book of matches, found in the kitchen by Fava.
3. (1) 38" by 48" piece of carpet, stained with an unknown liquid, found in the hallway by Conklin.

I collected each piece of evidence from the finder, marked it, packaged it, tagged it, and booked it at the El Fuego Police Department evidence room.

This example will accommodate any number of searchers and any number of evidence items. It can expand or contract as needed to establish good control of the evidence. The key to success with any type of evidence is to properly handle and care for it, and be able to account for it from the moment you seize it until you testify about it in court. (See Figures 5-7 and 5-8.)

SUMMARY

An investigator will meet many people during an inquiry who need to be identified in a report. Understanding what makes a person a victim, suspect, witness, reporting party, or "other" is key to categorizing that individual correctly. Knowing when to list someone as a felony suspect and the ramifications of doing so is vital to doing a good job.

Being able to help witnesses provide good descriptions of suspects is based on a couple of skills. First, the investigator must know what information is needed and in what order it will be broadcast to other investigators, and second, the investigator must be able to help the witness find references based on height and weight that help her or him recall what the person looked like. Using the formula for suspect descriptions when asking a witness to describe a suspect will more often than not result in a workable description.

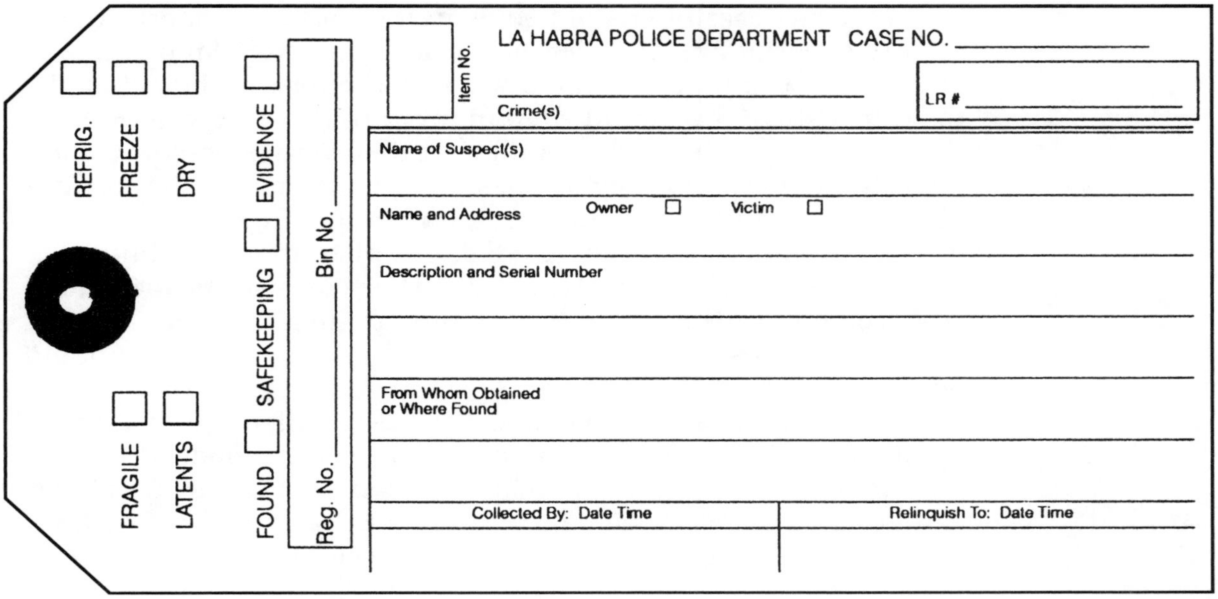

Figure 5-7 Front Side of Evidence Tag. (Courtesy of the La Habra Police Department)

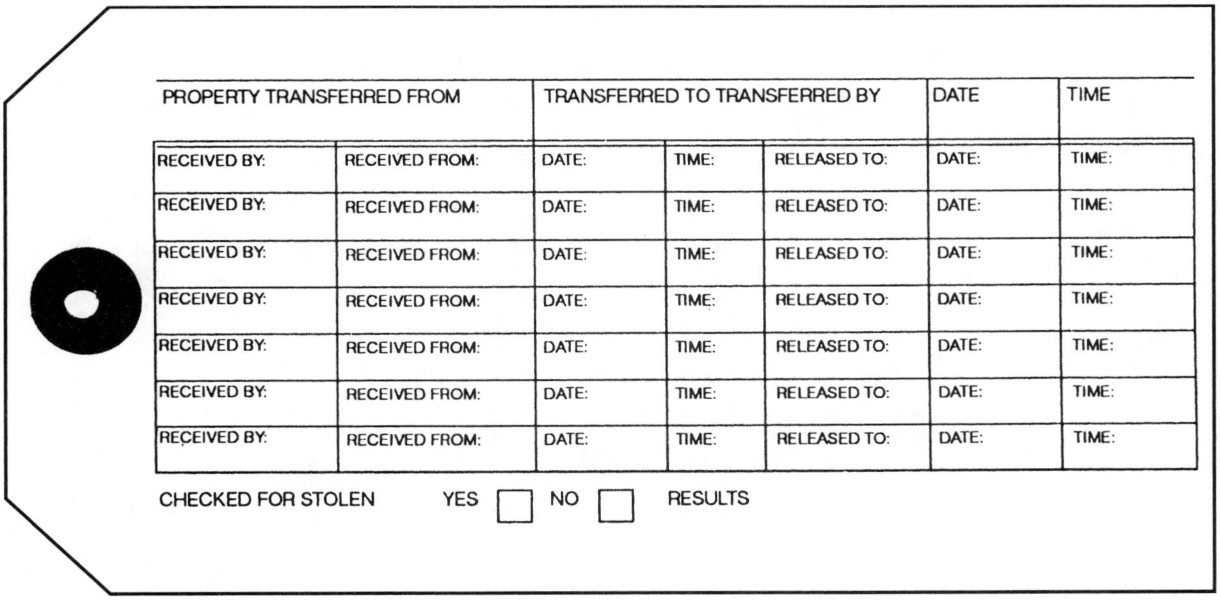

Figure 5-8 Reverse Side of Evidence Tag. (Courtesy of the La Habra Police Department)

Just as descriptions of people in a report are important, so too are descriptions of stolen or found property. A long and detailed description may not always be the best tool for this task. Investigators should keep in mind that the descriptions they use should be good enough that the average person could identify the property they are describing and that sketches and photos can be especially useful at times. Not only is a good description of the property needed, but also there are times when a value must be assigned. Whether you use the original cost, the fair market value, the victim appraisal, or the replacement cost method, be consistent and follow departmental policy.

All of the rules and guidelines for describing property also apply to the description of evidence with the additional requirement that a chain of custody must be established. Taking care of the evidence and ensuring that it is handled properly and accounted for at all times is a key to a successful investigation.

REVIEW

1. The people in a report are defined by the role each plays. List someone as a felony suspect only when there is probable cause to arrest him or her.

2. A witness is someone with useful information.

3. The name of everyone you talk to should be included in the report.

4. Property descriptions must pass the average person test.

5. Four methods of determining the value of stolen property are:

   Original cost method

   Fair market value method

   Victim appraisal method

   Replacement cost method

6. Evidence reports can be in free-flowing or fill-in-the-blanks form. A format for the free-flowing style is:

   GUNS

   MONEY

   DRUGS

   ITEMS WITH SERIAL NUMBERS

   ITEMS WITHOUT SERIAL NUMBERS

7. The chain of custody must be established.

EXERCISES

1. Measure and commit to memory the height of three reference points on your body.

2. Using these reference points, estimate the height and weight of five persons in your class. Were you close? Discuss why and how you can improve in this area.

3. Assume the classroom you are in was burglarized this past weekend and a color television set belonging to the school was taken along with a VCR belonging to the class instructor. Who are the victims? Explain your answer.

4. Write a description of the class instructor using the formula for describing persons.

5. Locate three articles in the classroom and describe each of them so that the descriptions pass the average person test.

6. Design a property report form.

7. Using the property report form, make an inventory of the appliances and electronic and entertainment equipment in your home.

8. Using the items in exercise 5, prepare an evidence report that includes a chain of custody. You are the finder of all three items, and you booked them at the Big Pine Police Department.

9. Using the same scenario as exercise 8, change the chain of custody to show three of your classmates as the finders and you as the person who collected and booked the items.

QUIZ

1. When describing stolen property in a report, how good should the description be?

    a. Completely accurate
    b. Only list the make, model, and serial number
    c. So it can pass the average person test
    d. As detailed as possible in thirty words or less

2. What is the definition of a victim?

3. What is the definition of a witness?

4. When should a person be named as a suspect in a felony crime report?

5. What is the order or formula for describing people in a report?

6. How should clothing be described?

7. If you have multiple witnesses and each provides a suspect description, when should the descriptions be combined into a composite description?

8. If a witness provides so much suspect information that it will not fit on the face page, where should it go?

9. Describe the original cost method of determining property value.

10. What is the chain of custody?

6

Crime Reports

KEY POINTS

In this chapter, we will cover all aspects of what is probably the most commonly written report in law enforcement, the crime report. We will begin with a definition of a **crime report** and explain why it is used, the parts of the report, and the single most important thing to be established in a crime report—the elements of the crime. We will also explore the purpose of the crime report face sheet and its two main uses and conclude with a discussion of the most common types of information needed to complete a crime report.

KEY TERMS

Corpus delicti

Crime report

Face sheet

Gather statistics

Investigative tool

Narrative section

Organize information

Solvability factors

Supplemental reports

The importance of the preliminary investigation cannot be overly emphasized because it forms the foundation for further investigative efforts and, for many crimes, represents the only investigation that is done. As such, it is imperative that the investigation be as thorough as possible and for the crime report to be clear, concise, accurate, and complete. The successful outcome of a case largely depends on the quality of the information collected during the preliminary investigation.

PURPOSE OF A CRIME REPORT

Of the hundreds of different reports used by the investigative organizations in this country, the most common is the crime report. A crime report can have many purposes depending on who is using it, but its application in the investigative process is well defined. First and foremost, the purpose of a crime report is to document that a crime has been reported. The most important thing to be established in a crime report is the **corpus delicti**, or elements of the crime. If the corpus is not established, there is no need for a crime report. This is not to say that if there is no crime, then there is no need for a report of some kind; rather, in these cases, the crime report is not the best way to document the incident. There is normally a great need to document the events an agency investigates, and most departments have an appropriate method to do so other than with a crime report. Although the requirements for completing a crime report vary from one law enforcement agency to another, a good rule of thumb to follow is that if the elements of a crime are present, a crime report should be completed. Some law enforcement personnel may challenge this point of view by saying that a crime report is not necessary if the victim does not desire prosecution; however, valuable information and evidence may be lost or become irretrievable if not documented or collected at the time.

Second, even though the victim may not be interested in moving forward with a prosecution at that point, he or she may change his or her mind. Third, the decision to prosecute someone does not rest with the officer or investigator but with the district attorney's office. Last but not least, documenting a crime when one is indicated allows an investigative agency to see the big picture with regard to activity and crime in the area. This knowledge allows the agency to plan deployment and resource strategies accordingly. This last point bears further note because almost without exception, law enforcement managers forecast deployment plans and staffing levels based in great part on the number of documented crimes occurring within certain time frames in a specific geographic area.

Completing the report as soon as possible eliminates the need for another officer to return later and do so. This wasteful practice of sending an investigator out to the scene a second time not only takes time and costs money but also lowers morale and fosters ill feelings among the officers having to do the work previously assigned to others.

Law enforcement agencies use crime reports for many purposes including the identification of suspects, listing stolen property, establishing methods of operation being used by area criminals, determining when crimes occurred in order to staff appropriately, documenting statistics for a multitude of reasons, and providing a way of justifying the arrest of those believed to have committed the crimes. The most important use of a crime report, however, is as an **investigative tool**.

Included in the responsibility law enforcement has to the community it serves is the role of helping those in need. Another major part of this responsibility is finding and bringing criminals to justice. It is difficult, if not impossible, to investigate these incidents of criminal wrongdoing without adequate crime reporting and an initial investigation.

Investigators who are charged with the task of completing crime reports must recognize that when they are assigned a radio call of a burglary or petty theft, they are not being assigned to go to the location and take a report; rather, they are being given the opportunity to investigate the incident and report what they learn. As discussed in an earlier chapter, an investigator is bound only by the law and his or her imagination and energy level.

Although the purpose and goals among the investigative agencies in the United States are similar, the manner and style of crime reporting can differ. Just as each agency has its own uniform for its investigators and color combinations for its patrol cars, each has its own method and configuration of crime report documentation.

COMPLETING CRIME REPORTS

Although there is a difference in the way these agencies complete crime reports and in the configuration of their crime report face sheets, the face sheets and accompanying free-flowing narrative sections are relatively simple in design and use. A crime report has two basic parts: the **face sheet**, which is generally a fill-in-the-blanks report, and the **narrative section** as described in chapter 4, which contains the investigation information. To simplify the task of completing the face sheet, it is helpful to understand its purpose. Despite the wide variety of configurations and boxes present on the thousands of face sheets used by this country's investigative agencies, the purpose of a crime report face sheet is twofold. One is to **organize information**, and the other is to **gather statistics**. (See Figure 6-1)

Once the investigator is familiar with the information that is requested on the face sheet, he or she can adjust the style of interviewing so that the questions asked of the victim or witness are in the same order they occur on the face sheet. This will reduce the time needed to fill in the blanks. In spite of the different crime

| ☐ NO PROSECUTION DESIRED
☐ TELEPHONE REPORT
☐ INSURANCE REPORT
☐ COURTESY REPORT
☐ DOMESTIC VIOLENCE
☐ CONFIDENTIAL SEX CRIME | **EL SEGUNDO POLICE DEPARTMENT**
348 MAIN STREET
EL SEGUNDO, CA 90245
310-322-9114
CRIME REPORT | A ☐ ACTIVE
S ☐ SUSPENDED
R ☐ RECORDS
C ☐ CLOSED
K ☐ COURTESY
U ☐ UNFOUNDED | CASE NUMBER

REFER OTHER RPTS |

CRIME

CODE SECTION	CRIME				UCR CODE	SECONDARY-COUNTS	OTHER-COUNTS
SPECIFIC LOCATION OF CRIME				OCCURRED ON/OR BETWEEN:	DATE	DAY	TIME
BUSINESS NAME		DATE RPT'D	TIME RPT'D	AND:	DATE	DAY	TIME

VICTIM

NAME (Last, First, Middle)		OCCUPATION		D.O.B.	AGE	SEX ☐ 1. M ☐ 2. F	RACE ☐ 1. WHT ☐ 2. HISP ☐ 3. BLK ☐ 5. CHI ☐ 7. FIL ☐ 9. P.ISL. ☐ 4. IND ☐ 6. JAP ☐ 8. OTH._____
RESIDENCE ADDRESS			CITY	ZIP CODE		RES. PHONE ( )	
BUSINESS NAME AND ADDRESS			CITY	ZIP CODE		BUS. PHONE ( )	

VICTIM(S) - WITNESS - RP

CODE	NAME (Last, First, Middle)	OCCUPATION		D.O.B.	AGE	SEX ☐ 1. M ☐ 2. F	RACE ☐ 1. WHT ☐ 2. HISP ☐ 3. BLK ☐ 5. CHI ☐ 7. FIL ☐ 9. P.ISL. ☐ 4. IND ☐ 6. JAP ☐ 8. OTH._____
RESIDENCE ADDRESS			CITY	ZIP CODE		RES. PHONE ( )	
BUSINESS NAME AND ADDRESS			CITY	ZIP CODE		BUS. PHONE ( )	
CODE	NAME (Last, First, Middle)	OCCUPATION		D.O.B.	AGE	SEX ☐ 1. M ☐ 2. F	RACE ☐ 1. WHT ☐ 2. HISP ☐ 3. BLK ☐ 5. CHI ☐ 7. FIL ☐ 9. P.ISL. ☐ 4. IND ☐ 6. JAP ☐ 8. OTH._____
RESIDENCE ADDRESS			CITY	ZIP CODE		RES. PHONE ( )	
BUSINESS NAME AND ADDRESS			CITY	ZIP CODE		BUS. PHONE ( )	
CODE	NAME (Last, First, Middle)	OCCUPATION		D.O.B.	AGE	SEX ☐ 1. M ☐ 2. F	RACE ☐ 1. WHT ☐ 2. HISP ☐ 3. BLK ☐ 5. CHI ☐ 7. FIL ☐ 9. P.ISL. ☐ 4. IND ☐ 6. JAP ☐ 8. OTH._____
RESIDENCE ADDRESS			CITY	ZIP CODE		RES. PHONE ( )	
BUSINESS ADDRESS			CITY	ZIP CODE		BUS. PHONE ( )	

VIC VEH

LICENSE #	STATE	YEAR	MAKE	MODEL	BODY STYLE ☐ 0 UNK ☐ 2 4-DR ☐ 4 P/U ☐ 6 VAN ☐ 8 RV ☐ 10 OTHER ☐ 1 2-DR ☐ 3 CONV ☐ 5 TRUCK ☐ 7 S/W ☐ 9 M/C _____
COLOR/COLOR		OTHER CHARACTERISTICS (i.e. T/C Damage, Unique Marks or Paint, etc.)		DISPOSITION OF VEHICLE	

FACTORS

☐ 1 THERE IS A WITNESS TO THE CRIME   SUSPECT PAGE ☐ YES ☐ NO
☐ 2 A SUSPECT WAS ARRESTED
☐ 3 A SUSPECT WAS NAMED
☐ 4 A SUSPECT CAN BE LOCATED
☐ 5 A SUSPECT CAN BE DESCRIBED
☐ 6 A SUSPECT CAN BE IDENTIFIED
☐ 7 A SUSPECT VEHICLE CAN BE IDENTIFIED
☐ 8 THERE IS IDENTIFIABLE STOLEN PROPERTY
☐ 9 THERE IS A SIGNIFIANT M.O.
☐ 10 SIGNIFICANT PHYSICAL EVIDENCE IS PRESENT
☐ 11 THERE IS A MAJOR INJURY/SEX CRIME INVOLVED
☐ 12 THERE IS A GOOD POSSIBILITY OF A SOLUTION
☐ 13 FURTHER INVESTIGATION NEEDED
☐ 14 CRIME IS GANG RELATED
☐ 15 HATE CRIME RELATED

EVIDENCE

☐ 0 NONE   ☐ 10 BLOOD
☐ 1 FINGERPRINTS   ☐ 11 URINE
☐ 2 TOOLS   ☐ 12 HAIR
☐ 3 TOOL MARKINGS   ☐ 13 FIREARMS
☐ 4 GLASS   ☐ 14 PHOTOGRAPHS
☐ 5 PAINT   ☐ 15 OTHER (DESCRIBE)
☐ 6 BULLET CASING   _____
☐ 7 BULLET   _____
☐ 8 RAPE KIT   _____
☐ 9 SEMEN

VICTIMS SIGNATURE	DATE	DETECTIVE ASSIGNED SIGNATURE	DATE

REPORTING OFFICER	ID#	DATE	REVIEWING SUPERVISOR	ID#	DATE

COPIES: ☐ CHIEF ☐ CII ☐ PATROL ☐ DB ☐ OTHER   ROUTED BY   ENTERED BY
TO: ☐ DMV ☐ CAU ☐ ABC (2 copies) ☐ DA _____

ESPD Form 3/97

Figure 6-1 Crime Report. (Courtesy of the El Segundo Police Department)

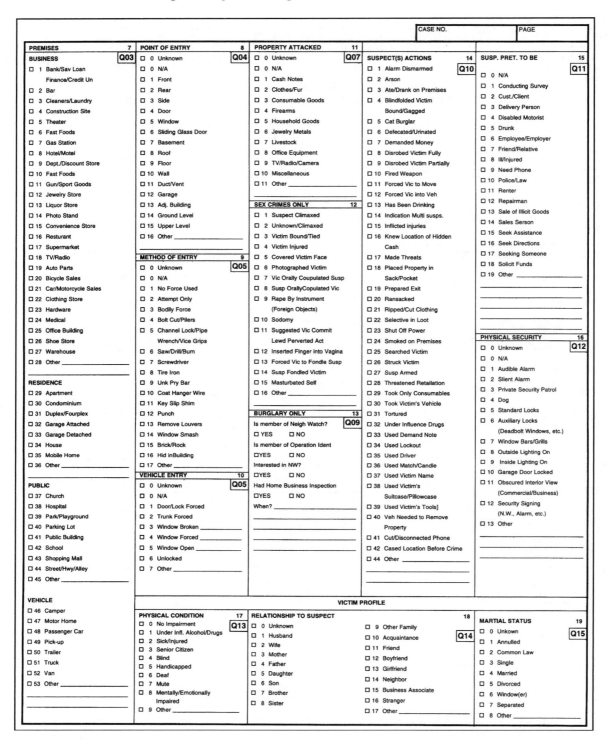

CASE NO.		PAGE	

PREMISES 7 | **POINT OF ENTRY** 8 | **PROPERTY ATTACKED** 11 | **SUSPECT(S) ACTIONS** 14 | **SUSP. PRET. TO BE** 15

BUSINESS Q03

☐ 1 Bank/Sav Loan Finance/Credit Un
☐ 2 Bar
☐ 3 Cleaners/Laundry
☐ 4 Construction Site
☐ 5 Theater
☐ 6 Fast Foods
☐ 7 Gas Station
☐ 8 Hotel/Motel
☐ 9 Dept./Discount Store
☐ 10 Fast Foods
☐ 11 Gun/Sport Goods
☐ 12 Jewelry Store
☐ 13 Liquor Store
☐ 14 Photo Stand
☐ 15 Convenience Store
☐ 16 Resturant
☐ 17 Supermarket
☐ 18 TV/Radio
☐ 19 Auto Parts
☐ 20 Bicycle Sales
☐ 21 Car/Motorcycle Sales
☐ 22 Clothing Store
☐ 23 Hardware
☐ 24 Medical
☐ 25 Office Building
☐ 26 Shoe Store
☐ 27 Warehouse
☐ 28 Other _____

RESIDENCE
☐ 29 Apartment
☐ 30 Condominium
☐ 31 Duplex/Fourplex
☐ 32 Garage Attached
☐ 33 Garage Detached
☐ 34 House
☐ 35 Mobile Home
☐ 36 Other _____

PUBLIC
☐ 37 Church
☐ 38 Hospital
☐ 39 Park/Playground
☐ 40 Parking Lot
☐ 41 Public Building
☐ 42 School
☐ 43 Shopping Mall
☐ 44 Street/Hwy/Alley
☐ 45 Other _____

VEHICLE
☐ 46 Camper
☐ 47 Motor Home
☐ 48 Passenger Car
☐ 49 Pick-up
☐ 50 Trailer
☐ 51 Truck
☐ 52 Van
☐ 53 Other _____
_____
_____

POINT OF ENTRY 8 Q04
☐ 0 Unknown
☐ 0 N/A
☐ 1 Front
☐ 2 Rear
☐ 3 Side
☐ 4 Door
☐ 5 Window
☐ 6 Sliding Glass Door
☐ 7 Basement
☐ 8 Roof
☐ 9 Floor
☐ 10 Wall
☐ 11 Duct/Vent
☐ 12 Garage
☐ 13 Adj. Building
☐ 14 Ground Level
☐ 15 Upper Level
☐ 16 Other _____

METHOD OF ENTRY 9 Q05
☐ 0 Unknown
☐ 0 N/A
☐ 1 No Force Used
☐ 2 Attempt Only
☐ 3 Bodily Force
☐ 4 Bolt Cut/Pliers
☐ 5 Channel Lock/Pipe Wrench/Vice Grips
☐ 6 Saw/Drill/Burn
☐ 7 Screwdriver
☐ 8 Tire Iron
☐ 9 Unk Pry Bar
☐ 10 Coat Hanger Wire
☐ 11 Key Slip Shim
☐ 12 Punch
☐ 13 Remove Louvers
☐ 14 Window Smash
☐ 15 Brick/Rock
☐ 16 Hid inBuilding
☐ 17 Other _____

VEHICLE ENTRY 10 Q05
☐ 0 Unknown
☐ 0 N/A
☐ 1 Door/Lock Forced
☐ 2 Trunk Forced
☐ 3 Window Broken _____
☐ 4 Window Forced
☐ 5 Window Open _____
☐ 6 Unlocked
☐ 7 Other _____

PROPERTY ATTACKED 11 Q07
☐ 0 Unknown
☐ 0 N/A
☐ 1 Cash Notes
☐ 2 Clothes/Fur
☐ 3 Consumable Goods
☐ 4 Firearms
☐ 5 Household Goods
☐ 6 Jewelry Metals
☐ 7 Livestock
☐ 8 Office Equipment
☐ 9 TV/Radio/Camera
☐ 10 Miscellaneous
☐ 11 Other _____

SEX CRIMES ONLY 12
☐ 1 Suspect Climaxed
☐ 2 Unknown/Climaxed
☐ 3 Victim Bound/Tied
☐ 4 Victim Injured
☐ 5 Covered Victim Face
☐ 6 Photographed Victim
☐ 7 Vic Orally Coupulated Susp
☐ 8 Susp OrallyCopulated Vic
☐ 9 Rape By Instrument (Foreign Objects)
☐ 10 Sodomy
☐ 11 Suggested Vic Commit Lewd Perverted Act
☐ 12 Inserted Finger into Vagina
☐ 13 Forced Vic to Fondle Susp
☐ 14 Susp Fondled Victim
☐ 15 Masturbated Self
☐ 16 Other _____

BURGLARY ONLY 13
Is member of Neigh Watch? Q09
☐ YES   ☐ NO
Is member of Operation Ident
☐ YES   ☐ NO
Interested in NW?
☐ YES   ☐ NO
Had Home Business Inspection
☐ YES   ☐ NO
When? _____
_____
_____
_____

SUSPECT(S) ACTIONS 14
☐ 1 Alarm Dismarmed Q10
☐ 2 Arson
☐ 3 Ate/Drank on Premises
☐ 4 Blindfolded Victim Bound/Gagged
☐ 5 Cat Burglar
☐ 6 Defecated/Urinated
☐ 7 Demanded Money
☐ 8 Disrobed Victim Fully
☐ 9 Disrobed Victim Partially
☐ 10 Fired Weapon
☐ 11 Forced Vic to Move
☐ 12 Forced Vic into Veh
☐ 13 Has Been Drinking
☐ 14 Indication Multi susps.
☐ 15 Inflicted injuries
☐ 16 Knew Location of Hidden Cash
☐ 17 Made Threats
☐ 18 Placed Property in Sack/Pocket
☐ 19 Prepared Exit
☐ 20 Ransacked
☐ 21 Ripped/Cut Clothing
☐ 22 Selective in Loot
☐ 23 Shut Off Power
☐ 24 Smoked on Premises
☐ 25 Searched Victim
☐ 26 Struck Victim
☐ 27 Susp Armed
☐ 28 Threatened Retaliation
☐ 29 Took Only Consumables
☐ 30 Took Victim's Vehicle
☐ 31 Tortured
☐ 32 Under Influence Drugs
☐ 33 Used Demand Note
☐ 34 Used Lockout
☐ 35 Used Driver
☐ 36 Used Match/Candle
☐ 37 Used Victim Name
☐ 38 Used Victim's Suitcase/Pillowcase
☐ 39 Used Victim's Tools]
☐ 40 Veh Needed to Remove Property
☐ 41 Cut/Disconnected Phone
☐ 42 Cased Location Before Crime
☐ 44 Other _____
_____
_____
_____

SUSP. PRET. TO BE 15 Q11
☐ 0 N/A
☐ 1 Conducting Survey
☐ 2 Cust./Client
☐ 3 Delivery Person
☐ 4 Disabled Motorist
☐ 5 Drunk
☐ 6 Employee/Employer
☐ 7 Friend/Relative
☐ 8 Ill/Injured
☐ 9 Need Phone
☐ 10 Police/Law
☐ 11 Renter
☐ 12 Repairman
☐ 13 Sale of Illicit Goods
☐ 14 Sales Serson
☐ 15 Seek Assistance
☐ 16 Seek Directions
☐ 17 Seeking Someone
☐ 18 Solicit Funds
☐ 19 Other _____
_____
_____
_____
_____
_____

PHYSICAL SECURITY 16 Q12
☐ 0 Unknown
☐ 0 N/A
☐ 1 Audible Alarm
☐ 2 Silent Alarm
☐ 3 Private Security Patrol
☐ 4 Dog
☐ 5 Standard Locks
☐ 6 Auxiliary Locks (Deadbolt Windows, etc.)
☐ 7 Window Bars/Grills
☐ 8 Outside Lighting On
☐ 9 Inside Lighting On
☐ 10 Garage Door Locked
☐ 11 Obscured Interior View (Commercial/Business)
☐ 12 Security Signing (N.W., Alarm, etc.)
☐ 13 Other _____
_____
_____
_____

VICTIM PROFILE

PHYSICAL CONDITION 17 Q13
☐ 0 No Impairment
☐ 1 Under Infl. Alcohol/Drugs
☐ 2 Sick/Injured
☐ 3 Senior Citizen
☐ 4 Blind
☐ 5 Handicapped
☐ 6 Deaf
☐ 7 Mute
☐ 8 Mentally/Emotionally Impaired
☐ 9 Other _____

RELATIONSHIP TO SUSPECT 18 Q14
☐ 0 Unknown
☐ 1 Husband
☐ 2 Wife
☐ 3 Mother
☐ 4 Father
☐ 5 Daughter
☐ 6 Son
☐ 7 Brother
☐ 8 Sister
☐ 9 Other Family
☐ 10 Acquaintance
☐ 11 Friend
☐ 12 Boyfriend
☐ 13 Girlfriend
☐ 14 Neighbor
☐ 15 Business Associate
☐ 16 Stranger
☐ 17 Other _____

MARTIAL STATUS 19 Q15
☐ 0 Unknown
☐ 1 Annulled
☐ 2 Common Law
☐ 3 Single
☐ 4 Married
☐ 5 Divorced
☐ 6 Window(er)
☐ 7 Separated
☐ 8 Other _____

Figure 6-1   (*Continued*)

report face sheets, most ask for similar information. Some of the commonly requested information include:

COPIES TO. This will generally be located in one of the corners of the face sheet and should be used to write the name of any investigator within the agency or the name of any outside agency that the writer wishes to receive a copy of the report.

CASE NUMBER. This refers to desk report and crime report numbers. These terms describe numbering systems used to file and index the documents and reports an investigative agency is involved with.

PREPARED BY. This is asking for the name, rank, and any badge or identification number used by the preparer.

OCCURRED ON. The date and time of occurrence and the day of the week are entered in this area. If the exact date or time cannot be established, the dates or times within which the crime took place should be listed.

CRIME. In the space calling for the crime, the investigator should place the code section and name of the crime. Lesser and included offenses should not be listed when perpetrated against a single victim. Generally, the most serious crime shown by the elements present is used, and only crimes substantiated in the narrative section of the report should be listed.

DATE AND TIME REPORTED. In this section, write the date and time the crime was first reported to your agency. This could be the time a complaint operator received the call or a victim or complainant in the field first spoke to you.

RD OR REPORTING DISTRICT. Enter the reporting district number where the crime occurred. Several reporting districts usually comprise an area or beat, and reporting districts are used to determine activity levels and calls for service in a geographical space.

LOCATION OF CRIME. In the block asking for the location of the occurrence or crime, pinpoint the location as closely as possible. In a report in which a business is listed as the victim, it is imperative that the business address be recorded. When the specific location is impossible to determine, use the vicinity of the crime. If the location is the same as that of the victim's address, you can generally write "same as above," as long as it is clear what you are referring to.

CSI. There is usually a box to check if a crime scene investigator or evidence technician examined the crime scene. It may also be necessary to write her or his name and identification number in the report.

SOLVABILITY FACTORS.   Sometimes referred to as "factors," these are general questions that the investigator can answer by checking a yes or no box for each question asked. Supervisors and follow-up investigators use **solvability factors** to prioritize cases with the greatest chance of being solved based on the information and evidence available. There are no hard and fast rules for selecting a yes or no answer, and no single answer is a determining factor in the decision to continue the investigation.

VICTIM'S NAME.   In this section, write the last name first, then the first name, followed by the full middle name. The last name might be followed by a comma or underscored, or both, when the name could be mistaken for a given name. If the report is to be typed, the last name might be in uppercase type to help distinguish it. When the victim has no middle name or initial, write "NMN," which means "no middle name." If the victim is a company, the company name should be written. If an ABC Oil Company service station is burglarized and money belonging to the company is taken, the victim is the ABC Oil Company. If a robbery is committed against the manager or any other employee of the same service station from which company money or property is taken, the firm is the victim and the employee would be a witness. If in addition to company money, the personal funds and property of an employee are taken, the employee would also be named as a victim. If more than one victim is involved in the same offense, the symbol V-1 should precede the name written in this section, and the names of other victims should be preceded by the symbols V-2, V-3, and so on. Additional victims should be listed in the narrative section and appropriately titled. (See Figure 6-2.)

VICTIM'S ADDRESS.   In this section, list the principal victim's address when the victim is an individual. If the victim is a business, list the business address.

RESIDENCE PHONE.   In the box asking for the residence phone, list the home phone number of the victim, or in the case of a business, the residence phone of the owner. Always precede the number with the area code.

BUSINESS PHONE.   In this section, write the telephone number at which the victim can be reached while at work. Always precede the number with the area code. If the victim is unemployed, write "NONE." In the event the victim is a place of business, write the telephone number of the business and write "DAY" in front of the daytime phone number.

VICTIM'S OCCUPATION.   In this box, write the victim's normal means of making a living. If the victim is unemployed, use the type of occupation the victim is normally employed in and write, for example, "UNEMPLOYED CARPENTER." When the victim is a business establishment, write the type of business that is

EL SEGUNDO POLICE DEPARTMENT	ADDITIONAL VICTIMS/WITNESSES	PAGE_____ OF _____

CASE NO

CRIME 1

CODE SECTION	CRIME	CLASSIFICATION	REFER OTHER REPORTS

LOCATION	RD.	DATE	TIME	SUPPL. ☐	INCIDENT NO.

WITNESS/RP/OR ADD. VICTIM 2

The following block repeats 8 times:

CODE	NAME (Last, First, Middle)	OCCUPATION	D.O.B.	AGE	SEX ☐ 1. M ☐ 2. F	RACE ☐ 1. WHT ☐ 2. HISP ☐ 3. BLK ☐ 5. CHI ☐ 7. FIL ☐ 9. P.ISL. ☐ 4. IND ☐ 6. JAP ☐ 8. OTH._____

RESIDENCE ADDRESS	CITY	ZIP CODE	RES. PHONE ( )

BUSINESS ADDRESS	CITY	ZIP CODE	BUS. PHONE ( )

REPORTING OFFICER	ID#	DATE	REVIEWED BY		ID#	DATE

COPIES: TO:	☐ CHIEF ☐ CII ☐ DMV ☐ CAU	☐ PATROL ☐ ABC (2 copies)	☐ DB ☐ DA	☐ OTHER AGENCY _____	ROUTED BY	ENTERED BY

ESPD Form #269 (Rev 5/97)

Figure 6-2 Additional Victims/Witnesses Form. (Courtesy of the El Segundo Police Department)

conducted there unless the name of the business precludes the necessity of further explanation, for example, "JOE'S BAR." If the victim is a student, write the name of the school attended and the city where the school is located. For example, Student, City High School, San Diego.

DOB.   Write the victim's date of birth when available and applicable.

PERSON REPORTING OFFENSE.   In many instances, this will be the same individual previously listed as the victim. If such is the case, write the victim's last and first names here. However, if the person reporting the offense is someone other than the victim, this should be recorded. If the victim is a business and the reporting party is an employee, his or her relationship with the business should be described.

PERSON WHO DISCOVERED THE CRIME.   Many times this will be the same person as the victim or the person reporting. If so, the details for this box have already been obtained. If, however, the person reporting the offense is someone other than those already mentioned, this should be recorded. If the victim is a business and the person discovering the crime is an employee, this relationship with the business should be shown.

M.O. SECTION.   The M.O., or modus operandi, section is generally completed for robberies, burglaries, and aggravated sex cases as well as for other crimes when the investigator believes the information would be helpful in solving the crime. This is the area to write the basic method used by the suspect to commit the crime, along with any unusual things the suspect did during its commission.

CHARACTERISTICS OF PREMISES OR AREA.   This is where the characteristics or type of place where the crime was committed is written. This information, in many cases, will be a description of the size, type, area, or characteristics of the neighborhood. Examples are "one-story, five-room, single-family residence on a corner lot" and "mobile home in a largely unattended mobile home park."

MOTIVE—TYPE OF PROPERTY TAKEN OR OTHER REASON FOR OFFENSE.   Generally, the class or type of property taken or the motive or reason why the offense was committed is written in this area. In crimes in which property has been taken, the motive will usually be the type of property taken in support of personal gain. The specific type of property that was taken or attempted to be taken should be listed. This might include money and jewelry, women's clothing, cigarettes, and narcotics. The detailed description of the property and any serial numbers should appear in the property of loss section. In other types of crimes, the motive might be revenge, insurance settlement, concealment of crime, sexual gratification, ransom, or, in narcotics cases, money from the sale of narcotics or the effects resulting from their use. In some cases, the initial crime may lead to a second offense, for example, a case in which a homicide is committed during a robbery or attempted rape. In such cases, the motive would be robbery or rape.

VICTIM'S ACTIVITY JUST PRIOR TO AND/OR DURING THE OFFENSE. The victim's activity at these times may characterize the kind of person the offender selected as a victim. In rape cases, the victim's activity just prior to the attack might be "waiting at bus stop," "doing laundry in a laundromat," "entering car in parking lot," or "in bed asleep." In a robbery case, the victim's activity just prior to the offense might be "walking down street," "waiting on customers," or "closing store." With a burglary case, the victim's activity during the offense might be "on vacation," "attending a funeral," or "home in bed." When the victim is a business, the victim's activity is likely to be either "open for business" or "closed for business." When "open for business" is used, include the natural activity of the attendant or employee just prior to or during the attack.

DESCRIBE WEAPON, INSTRUMENT, EQUIPMENT, TRICK, DEVICE, OR FORCE USED. For crimes against property, list the type of tool used and its size if this can be determined. If tools are not used, write what was used, such as, hands, feet, voice, etc. For crimes against persons, write as complete a description of the weapons used as possible. If force was used, describe the force, for example, "knocked to ground," "kicked," "hit with fist," or "threatened with unknown-type liquid."

WHAT SUSPECT SAID. If possible, write the exact words used by the suspect. Pay particular attention to any mispronunciations, unusual words, peculiar expressions, accents, or dialects. Many times the way the suspect said something is just as important as what was said. Try to explain this if it is at all possible.

TRADEMARK OR OTHER DISTINCTIVE ACTION OF SUSPECT. Any action by the suspect in preparation for the crime, flight from the scene, or disposition of the proceeds of the crime that has not been recorded in any other category of the modus operandi should be written here. The act may be necessary for the completion of the crime but frequently is not. Preparations for the crime, as well as precautions to avoid apprehension or detection, may be necessary but are not included elsewhere in the report. Examples are "cased storeroom the day before," "wiped off fingerprints," or "closed venetian blinds but turned one slat to provide view of front entrance." Unnecessary acts are "eats food," "leaves note," or "plays stereo." The number of bizarre acts a suspect may do is unlimited.

ADDITIONAL VICTIMS. List additional victims and include all of the information you did for victim #1 such as name, address, residence phone, business phone, occupation, and date of birth. Identify additional victims as V-2, V-3, and so on.

WITNESSES/SUSPECTS. When there is more than one witness or suspect, number them using the same guidelines as for additional victims. Include as much information and as complete a description as possible. When a suspect is in custody, say so by writing "IN CUSTODY" followed by the name of the jail. Generally, suspects should be listed by name and description only when they could be arrested or a complaint charging them with the offense could be issued. Otherwise, list them in the narrative section of the report. Suspects' descriptions should be written separately in the event different witnesses give different descriptions. If something is not known about a person's description, leave it out, and never combine suspect descriptions into one.

VEHICLE USED BY SUSPECT. Write any information available as to the transportation used by the suspect. This may range from a complete description of an automobile to a partial one based on the size and number of tire tracks seen at the crime scene. When you cannot establish that a suspect used a vehicle, write "NONE SEEN OR HEARD."

PROPERTY. Each item should be numbered for easy indexing, and the quantity of each type of item stolen should also be shown. Following this, write a complete description of the property including the make, model, serial number, size, value, and any other identifiable marks or characteristics. Remember the average person test. A total value for the stolen property is usually called for in this area as well.

EVIDENCE. If the investigating officer or someone other than a crime scene investigator collects the evidence, the person responsible for the chain of custody should be listed on the report in this area. If a crime scene investigator collects the evidence, he or she will most likely prepare a report using the same case number as your investigation for easy reference. If this is the case, write "SEE CRIME SCENE INVESTIGATOR REPORT."

INJURIES. Describe who received injuries and their seriousness.

SUPPLEMENTAL REPORTS

Field officers will probably complete more crime reports than any other kind of report because of the large number of calls for service in which a crime is reported. Not far behind in volume for the field officer will be opportunities to complete supplemental reports.

A **supplemental report** is a general catchall for the numerous pieces of information that find their way into a law enforcement agency. Field officers will have numerous dealings with people who are reporting additional facts about recent investigations, as well as about cases that could be weeks or months old. Some agencies will have well-developed policies describing when and how supplemental reports should be completed, and others will leave the matter up to the discretion of the involved investigator.

Some agencies will have preprinted forms that function much like a crime report face sheet. These forms will have a limited fill-in-the-blanks section and a larger narrative area. Other agencies will use a blank form and require the investigator to show the proper information. No matter the style of the supplemental report, the investigator should include certain things. The most important is the case number, which will help connect the new information to the existing case file. If the case number is not known, a new one would be assigned. In either case, this will allow the case to be tracked effectively. The investigator should then follow the rules of narrative writing and ensure that all of the information is reported. Remember that even though something seems insignificant to you at the time, it may be very important to the detective handling the case and eliminate a blank spot in the big picture. Supplemental reports are important and should be given the same care, thought, and attention as all other kinds of reports.

Your reputation may rest on a report you write. Many people will read and evaluate your work, and you may have to base your testimony on your report as well. There may be times when information is limited, but you must make every reasonable attempt to complete the report. It may be possible that the best opportunity for solving a crime rests with the first investigator at the scene when the evidence is fresh and witnesses are more likely to be present and remember what happened. Some consider radio calls of a theft or a burglary as a task for which they have to take a report. In reality, each of these calls for service is an opportunity to investigate a crime and perhaps solve it. Regardless of your rank, assignment, or tenure, if you are assigned to conduct a lawful search for a thing or person and the goal is to find the truth, you are an investigator.

SUMMARY

One of the more common investigative duties is the completion of crime reports. The key to this is establishing that a crime occurred and including the elements, or corpus delicti, in the report. Crime reports usually have two distinct parts, the face sheet, the purpose of which is to organize information and gather statistics, and the narrative section, in which the investigator tells the story of what happened, using the rules of narrative writing. Other keys to a good-quality crime report are the identification and the description of the suspect's method of operation, which can often tie a captured suspect to other unsolved crimes.

REVIEW

1. Crime reports must establish the corpus delicti.
2. The purpose of a crime report face sheet is to organize information and gather statistics.
3. The two parts to a crime report are a fill-in-the-blanks face sheet and a free-flowing narrative.
4. Start the narrative with the date, time, and how you got involved.
5. Solvability factors help prioritize follow-up investigation.
6. Supplemental reports have many uses.

EXERCISES

1. Visit three police departments in your area and get copies of the crime report face sheets used by them. What are the similarities?
2. Are there any significant differences among the reports? How will this affect the preliminary investigation of an incident?
3. View a law enforcement action show and take notes on the calls for service. Using a face sheet, complete a crime report.
4. Divide the class into two groups, with one group playing the victim and the other the officer. Using scenarios supplied by the instructor, conduct a preliminary investigation and complete a crime report.
5. Review the crime report, identify the areas needing improvement, and rewrite the crime report using the rules of narrative writing.
6. Crime report exercise. Use the following fact pattern and prepare an appropriate crime report.

FACTS: You are a patrol officer for the El Fuego Police Department working Area 12. Your back up officer is JEFF GUIDRY. Use your state and ZIP code for all address needs.
CRIME: Vandalism RD 822 BEAT 12
LOCATION: 6221 Wood Street, El Fuego
OCCURRENCE DATE: November 15, 2006
REPORTED DATE: November 15, 2006 1445 hours
RP: Meyers, Tammy Lynn DOB 10/13/84
DISCOVERER: Runyan, John DOB 4-30-85
VICTIM: Same as the RP
M.O.   There is no M.O. For this report
SOLVABILITY   FACTORS: Determine this using the fact pattern
PROPERTY DAMAGE: $150 to repair the window

DETAILS: You are on patrol and receive the following radio call, "Unit 12, see the woman at 6221 Wood Street regarding a vandalism in progress. Reporting party is Tammy Meyers and her boyfriend is in foot pursuit northbound on Main Street. Suspect is a male, white with blond hair wearing a red shirt and tan pants." As you near the intersection of Fanwood and Ash, which is 3 blocks from the victim's residence, you observe a young blond man in a red shirt running west on Fanwood. You also observe another man in a green shirt chasing him. The second man is about 25 yards behind the guy in the red shirt. You order both of them to stop and they readily comply. Both are breathing very hard, and both are sweating. You call dispatch and advise them of your actions and shortly thereafter hear Officer Guidry on the radio. Officer Guidry is at the victim's house and advises you that the vandalism involves a brick being thrown through a window. Officer Guidry advises that a suspect in a red shirt threw a brick through the living room window and that the victim's current boyfriend John Runyan was chasing the suspect. Officer Guidry also advises you that the suspect is named David Lockwood and is a former boyfriend of the victim.

You observe some red brick dust on the suspect's left hand and on the left side of his tan pants. The man who was chasing the suspect states that he is John Runyan and that he saw the suspect throw a brick through the living room window. He continues stating that he chased him from the house and never lost sight of him. He also states that the suspect used to date his current girlfriend and the suspect is pissed off that she no longer wants to go out with him. When you ask the suspect his name, he states, "I really am sorry, but I still hate her." Officer Guidry radios you and broadcasts that he has recovered a brick from the victim's living room. Based on this, you place the suspect under arrest and find an identification card in his wallet in the name of Dave Lockwood. Officer Jim Barnes transports him to jail.

You then respond to the victim's house to talk with Officer Guidry and the victim so that you can take a report. On arrival, you contact the victim who states that she used to go out with the suspect but broke up with him in August because he was so weird. You observe the broken window and glass on the floor of the living room. You complete your work and get the brick from Officer Guidry along with a Polaroid photo he took of the broken window.

Victim: Tammy Lynn Meyers DOB 10-13-84

6221 Wood Street, El Fuego

Home Phone (555) 613-2664

Work Phone (555) 281-1624

Meyers is a waitress at Joe's Burgers, 8193 West Street, El Fuego.

Witness: John David Runyan DOB 4-30-85

317 Harbor #18, El Fuego

Home phone (555) 235-7184

Work Phone (555) 281-1624

Runyan is a bartender at Joe's Burgers, 8193 West Street, El Fuego.

Property loss: $150 to replace the window

Suspect: David Shelton Lockwood DOB 1-6-86

210 Trinity Road, El Fuego

Home Phone (555) 641-4545

Work Phone: None

Unemployed

---

7. Crime report exercise: Use the following fact pattern and prepare an appropriate crime report using the rules of narrative writing. You are a patrol officer for the El Fuego Police Department working day shift in Area 6. Your radio call sign is "6ADAM1."

Date: September 15, 2006 At 1030 hours

Solvability Factors: Use the fact pattern to determine

DETAILS: You are on routine patrol and get the following radio call, "6ADAM1 see the man at 161 Nipomo regarding a theft—the reporting party is Bruce Berger." You respond to the victim's house and observe a man standing in his driveway next to a late model red Cadillac. As you question the man, you identify him as the victim, Bruce Berger. The victim tells you that he left his bicycle on the front porch last night around 6 P.M. and when he got up this morning, the bike was gone. The victim is really upset because the bike was an old Schwinn beach cruiser that he had owned for 20 years. The victim says he has a serial number somewhere and that the bike was yellow. The victim states to you that he did not hear anything during the night and that no one has permission to take the bike. After you take the victim's statement, you walk over to the neighbor's house across the street to see if the people living there saw anything. No one answers the door at 170 Nipomo and the house is quiet. You check back with the victim who gives you a picture of his bike and the serial number FY32164. He also states that the bike cost him $120 when it was new, but is worth twice that now because it is a classic bike. Just before you leave, you radio the dispatcher and get a case number of 06-31412 and put out a description of the stolen bike over the radio. You are familiar with the area and know that a lot of bicycles are stolen. Around 11:15 A.M., you return to routine patrol.

Victim: Bruce Berger DOB 3-12-54

161 Nipomo, El Fuego

Home Phone (555) 533-1616

Work Phone (555) 614-3200

Business: owner of Good Grinds Coffee Shop, 4128 Orange Street, El Fuego

8. Crime report exercise. Review the following crime report and identify any problems. Rewrite the report using the rules of narrative writing.

CRIME: Vandalism RD: 637 Beat: 2
LOCATION: 187 Van Ness Circle, El Fuego, CA
OCCURRENCE DATE: 7-6-07 to 7-7-07 between 2300-0530
REPORTED DATE: 7-16-07 at 1625 hours
REPORTING PERSON: Cleworth, Frank DOB 12-3-34

187 Van Ness Cr. #19C

El Fuego, CA

Home Phone (800) 555-6826

Work Phone (800) 555-6826

DISCOVERER: Same as above
VICTIM: Clear Picture Cable

154 Roswell, El Fuego

(800) 555-4910

THERE IS NO M.O. SECTION FOR THIS REPORT

SOLVABILITY FACTORS
     [Y]   IS THERE A SUSPECT?
     [Y]   IS THIS A CONTINUING PROBLEM?

PROPERTY DAMAGE: 2 1/2 feet of television cable, nfd; value $50.

DETAILS:

At the above indicated date and time, I was dispatched to the above location. Upon my arrival at the Van Ness address, I contacted the R/P, Mr. Cleworth.

Mr. Cleworth stated to me that the gentleman who lives in the apartment above him, Mr. Paul Johnson (unknown spelling), who resides at 187 Van Ness, apt. #19G, has been in an ongoing dispute and argument with the R/P.

He stated that on Saturday evening, 7-6-07, at approximately 2300 hours, he terminated watching television for the evening and retired. The very next morning, on Sunday, 7-7-07, at approximately

0530 hours, he awoke and responded to the living room of his condominium. There, he turned on the television, which was functioning, but was unable to receive any picture. He immediately contacted the Clear Picture Cable people that same day.
Mr. Cleworth indicated that Clear Picture Cable driver Paul V. S., #27, responded to his residence. There, he checked the cable.

It should be noted that there is an ordinance in the C.C.R.'s of the condominium complex that requires all cables to either be underground or concealed within the building. As a result, the television cable system that Mr. Cleworth ordered had to be installed through the attic of the two-story condominium building. From the attic, the cable was then run down along the inside of the exterior laundry room attached to the balcony of apartment #19G belonging to Mr. Johnson; from there, through the floor, into the laundry room located on the ground level below Mr. Johnson's condo belonging to Mr. Cleworth, and then from the laundry room into the living room of Mr. Cleworth's condominium located directly below that of Mr. Johnson.

Mr. Cleworth indicated that the cable television installer was unable to locate any difficulty with the cable on the ground level and that he was unable to locate any difficulty with the cable leading to the building, but was unable to gain access to Mr. Johnson's condominium at that time.

The same Clear Picture Cable representative returned to the location on 7-10-07. There, with the consent of the manager of the condominium complex, Mr. Joe Willy (unknown residence number, business phone number: 555-5475), the Clear Picture Cable representative, Paul V. S., #27, and the manager, Mr. Willy, climbed the ladder onto the balcony of condominium #19G belonging to Mr. Johnson. There, they opened the door leading to the laundry room located on the balcony and examined the interior. There, inside Mr. Johnson's laundry room, the installer discovered a 2 1/2 feet piece of the television cable to be cut away from the cable leading from the attic above, down to Mr. Cleworth's condominium. The installer subsequently repaired the splice, advised Mr. Cleworth of what he had found, and left the location.

This reporting officer did not see the damage indicated by Mr. Cleworth. Mr. Cleworth also indicated that the repairs were at the expense of the victim, Clear Picture Cable, who was at this time unavailable for comment.

Mr. Willy, the manager of the complex, was also unavailable for comment.

QUIZ

1. When should a crime report be completed?

2. What is the most important thing to be established in a crime report?

3. What is the most important use of a crime report?

4. What are the two basic parts of a crime report?

5. Why is the phrase "take a report" a misnomer?

6. What are solvability factors?

7. What is the purpose of a crime report face sheet?

8. In what time frame should a crime report be completed?

9. What is the minimum length of a crime report narrative?

10. How should the narrative of a crime report begin?

7

Arrest Reports

KEY POINTS

Some of the investigations you conduct will include the arrest of one or more suspects. In this chapter, we will cover the area of arrest reports, including what is needed to make an arrest and when to document the action you have taken. We will also explore the types of arrest report formats from the fill-in-the-blanks kind to the free-flowing narrative type that uses a functional format that can be adapted to your specific situation. We conclude the chapter with a brief discussion of the arresting officer's obligation in reporting what has occurred.

KEY TERMS

Face sheet

Fill-in-the-blanks form

Narrative

Probable cause

Report formats

Generally speaking, an arrest report is necessary whenever someone is taken into custody by a law enforcement agency as the result of a criminal investigation or in response to a citizen's request for a private person's arrest. Although fact patterns indicating an arrest are infinite, there are four basic reasons an arrest is made:

1. A police officer sees a misdemeanor crime occur.
2. A police officer believes a felony has been committed by a particular person.
3. A citizen makes a private person's arrest.
4. An arrest warrant exists for a person.

DOCUMENTING THE ARREST

Whichever one of these circumstances is the basis for the arrest, the responsibility for documenting it rests with the investigator who detains the suspect. His or her report should clearly show the circumstances of what happened and include the facts that led to the decision to make the arrest. These facts and circumstances combine to form the **probable cause**, which is the basis for the arrest. In criminal cases in which an arrest is made, probable cause is a reasonable belief that the person being arrested committed the act he or she is being charged with. The facts and circumstances for each arrest will vary, and as such, the events leading up to the decision to make an arrest will be different. This is why it is so important to include as much detail as possible when describing probable cause. Without exception, the single most important thing that must be established in an arrest report is probable cause.

ARREST REPORT STYLES

Depending on the circumstances and the length of the investigation, an arrest report might take one of several forms. Some agencies use an arrest report face sheet that is similar in appearance and function to a crime report face sheet in that it helps to organize information and allows for the easy gathering of statistical information. This **fill-in-the-blanks form** also requires a narrative section to be completed. Other agencies use a totally narrative style for arrest reports and gather statistics from the booking sheet. (See Figure 7-1.)

Whether the combination face sheet and narrative style or the all-narrative style is used, the key to preparing a complete, clear, concise, and accurate arrest report is to write a narrative section that explains what happened. You must explain who did what, how it happened, and where and when it took place. The rules of narrative writing described in chapter 4 apply to arrest report narratives as well and allow them to be completed quickly, professionally, and with consistency. The writer of an arrest report

LA HABRA POLICE DEPARTMENT DUI FIELD INTERVIEW REPORT		CASE NUMBER	
		PAGE	OF

DRIVER'S LAST NAME	DRIVERS LICENSE NUMBER	STATE	CLASS	STATUS	DATE / TIME OF FIELD INTERVIEW

PASSENGER NAME/DOB/ADDRESS/PHONE

VEH	YEAR	MAKE	MODEL	STYLE	COLOR	LICENSE PLATE #	STATE	ACCIDENT INVOLVED? ☐ YES ☐ NO

LOCATION OF :   ☐ TRAFFIC STOP   ☐ TRAFFIC COLLISION   R/O NAME & ADDRESS:

1. Do you know of anything mechanically wrong with your vehicle?   ☐ YES   ☐ NO   If yes, describe: _____

2. Are you sick or injured?   ☐ YES   ☐ NO   If yes, describe: _____

3. What time is it? _____ AM/PM   3a. Actual time? _____ Hrs.

4. Are you diabetic or epileptic?   ☐ YES   ☐ NO   If yes, describe: _____

5. Do you take insulin pills or injections?   ☐ YES   ☐ NO   If yes, describe: _____

6. Do you have any physical defects?   ☐ YES   ☐ NO   If yes, describe: _____

7. When did you last sleep? _____ AM/PM   7a. How long? _____

8. When did you last eat? _____   8a. Describe meal: _____

9. Have you bumped your head recently?   ☐ YES   ☐ NO   If yes, describe: _____

10. (If driving not observed) Were you driving the vehicle?   ☐ YES   ☐ NO

11. How long have you been driving today/tonight? _____

12. Where did you start driving? _____

13. Where were you going? _____

14. Where are you now? _____

15. What have you been drinking? _____

16. How much have you been drinking? _____

17. What time did you start drinking? _____

18. What time did you stop drinking? _____

19. Where were you drinking? _____

20. With whom were you drinking? _____

21. Do you feel any effects from the drinks?   ☐ YES   ☐ NO   If yes, describe: _____

22. Are you currently under the care of a Doctor or Den tist?   ☐ YES   ☐ NO   If yes, name and adress: _____

23. Have you had any recent surgery?   ☐ YES   ☐ NO   If yes, describe: _____

24. Have you taken any medicine?   ☐ YES   ☐ NO   If yes, describe: _____

25. What dosage? _____

26. What was the time of the last dose taken? _____

27. Have you had anything to drink in the last hour?   ☐ YES   ☐ NO   If yes, describe: _____

DESCRIBE TEST LOCATION: (WEATHER CONDITIONS, LIGHTING, GRADE ETC)

INVESTIGATION NOTES

REPORTING OFFICER / ID	WITNESSING OFFICER / ID	APPROVED BY

Figure 7-1   DUI Field Report. (Courtesy of the La Habra Police Department)

LA HABRA POLICE DEPARTMENT
DUI FIELD BALANCE TEST

CASE NUMBER:

DATE/TIME OF F.B.T.

BREATH	EYES	SPEECH	COORDINATION	APPEARANCE	BEHAVIOR	CLOTHING

☐ ALCOHOLIC
☐ MARIJUANA
☐ OTHER/PCP
☐ OTHER___
———
☐ STRONG
☐ MODERATE
☐ WEAK
☐ NONE
	☐ BLOODSHOT
	☐ WATERY
	☐ DROOPY
	☐ CLEAR
	
	PUPIL SIZE:
	☐ CONSTRICT
	☐ DILATED
	
	☐ OTHER_____
		☐ SLURRED
		☐ LOUD
		☐ SOFT/QUIET
		☐ MUMBLED
		☐ SPITTING
		☐ RAPID
		☐ TALKATIVE
		☐ SURE/
		☐ CORREC T
		☐ OTHER ___
			☐ UNSTEADY/
			   GAIT
			☐ USED/
			   SUPPORT
			☐ STAGGERED
			☐ FELL
			☐ FUMBLED ID
			☐ LAX FACE/
			   JAW
			☐ OTHER
				☐ UNKEMPT
				☐ VOMITUS
				☐ HAIR MESSED
				☐ DRY MOUTH
				☐ NORMAL
				
				☐ OTHER_____
				———
				———
					☐ ANGER
					☐ LAUGHING
					☐ CRYING
					☐ BELLIGERENT
					☐ AGGRESSIVE
					☐ ARROGANT
					☐ REMORSEFUL
					☐ INDIFFERENT
					☐ FLUCTUATING
					☐ COOPERATIVE
						☐ UNBUTTONED
						☐ UNZIPPED
						☐ DISHEVELED
						☐ URINE
						☐ NEAT
						☐ STAINED/
						   DIRTY
						DESCRIBE: _____
						_____
						_____

GLASSES CONTACTS	☐ YES ☐ NO	SHOES WORN DURING TEST	☐ YES ☐ NO	DESCRIBE SHOES:	NYSTAGMUS TEST GIVEN	☐ YES ☐ NO

FIELD BALANCE TESTS

ALPHABET TEST	MODIFIED POSITION OF ATTENTION	ONE LEG STANCE
☐ SURE, CORRECT ☐ SLOW, HESITANT ☐ SLOW, DELIBERATE ☐ STARTED OVER____TIMES ☐ SLURRED ☐ NOT UNDERSTANDABLE ☐ SKIPPING LETTERS A B C D E F G H I J K L M N O P Q R S T U V W X Y Z A B C D E F G H I J K L M N O P Q R S T U V W X Y Z	☐ SURE ☐ HESITANT ☐ USED ARMS FOR BALANCE ☐ LOST BALANCE ☐ FALLING ☐ OTHER_____ ☐ LATERAL SWAY_____" EXPLAIN: _____	LEFT ☐☐☐☐☐ SURE ☐ HESITANT ☐ USED ARMS FOR BALANCE ☐ LOST BALANCE ☐ FALLING ☐ OTHER ☐☐☐☐☐ RIGHT EXPLAIN: _____

FINGER TO NOSE TEST	HEEL TO TOE WALK
X = RIGHT 0 = LEFT LEFT ☐☐ SURE HESITANT SLOW/SEARCHING ☐☐ RIGHT ☐ TURNING HEAD ☐ TO MEET FINGER EXPLAIN:	X = RIGHT 0 = LEFT ———————————→ ←——————————— EXPLAIN:

ABILITY TO READ AND FOLLOW SIMPLE DIRECTIONS	( ) GOOD RETENTION & QUICK RESPONSE TO INSTRUCTIONS ( ) POOR RETENTION AND RESPONSE ( ) ATTEMPTS TESTS BEFORE / DURING INSTRUCTIONS	( ) INTERRUPTING ( ) EVASIVE QUESTIONS ( ) FAIR RETENTION AND RESPONSE

13353 CVC	BLOOD TEST	BREATH TEST	URINE TEST
13353 CVC ADVISAL BY OFFICER _____ (NAME/ID) ( ) ENGLISH ( ) REFUSAL ( ) SPANISH	SAMPLE TIME:_____HRS. TEST LOCATION:_____ LVN NAME:_____ VIAL #:_____	OPERATOR NAME _____ TEST LOCATION _____ RESULT #1 _____% BAC RESULT #2 _____% BAC	SAMPLE TIMES_____HRS ____HRS TEST LOCATION _____ SAMPLE ID# _____ OFFICER NAME / ID _____

REPORTING OFFICER/ / ID	ASSISTING OFFICER / ID	APPROVED BY / DATE

LHPD 91-11   REVISED 8/91

Figure 7-1   *(Continued)*

will soon realize that the narrative of a crime report and the narrative of an arrest report follow the same rules. They are written in the first person; past tense; active voice; in chronological order beginning with the date, time, and how you got involved; and use short, clear, concise, and concrete words.

COMPLETING THE ARREST REPORT

When completing an arrest report that uses a **face sheet**, begin by filling in the boxes as requested. Then begin the **narrative** with the date, time, and how you got involved, for example:

On 9-1-2006 at 2145 hours, I received a radio call of a prowler at 6101 Pine Street.
On 7-4-2006 about 1330 hours, I saw Brown, who was driving west on Pacific, fail to stop for the stop sign at Main Street.
On 12-19-2006 at 1720 hours, I was driving through the Del Lago Plaza parking lot when Brown hailed me and said her husband was chasing a purse snatcher. (See Figure 7-2.)

REPORT FORMATS

When no face sheet is used, the arresting officer must overcome the problems of creating a workable format and beginning the report. One **report format** that works well is to list pertinent headings of ARRESTED, CHARGE, LOCATION, DATE AND TIME, OFFICER, and DETAILS at the top left side of the page, complete the information based on the fact pattern, and begin the narrative. These headings, when set up properly, allow for multiple suspects to be listed while still organizing the information in a workable way. For example, the following is a suggested format when one suspect is arrested. (See Figure 7-3.)

ARRESTED:
CHARGE:
LOCATION:
DATE AND TIME:
OFFICER:
DETAILS:

The suggested format when two or more suspects are arrested is shown below. (See Figure 7-4.)

ARRESTED: 1.
CHARGE:
ARRESTED: 2.
CHARGE:

EL SEGUNDO POLICE DEPARTMENT	CASE NO.	PAGE
SUPPLEMENTAL/NARRATIVE (Check one) □ SUPPLEMENTAL REPORT □ NARRATIVE CONTINUATION		CA0192300

CODE SECTION	CRIME	VICTIM'S NAME (FIRM IF BUSINESS)

PREPARED BY				REVIEWED BY		
NAME	I.D. NUMBER	MO. DAY YR.	NAME		MO. DAY YR.

P-16 REV 7-98

Figure 7-2 Supplemental Narrative. (Courtesy of the El Segundo Police Department)

EL FUEGO POLICE DEPARTMENT

ARREST REPORT   98-6562110

ARRESTED: Rabbit, Donald DOB 6-18-52

CHARGE: Warrant F246381

                 211 PC, Robbery

LOCATION: Main and Walnut, El Fuego

DATE and TIME: 12-16-98 1830 hours

OFFICER: J. Fava #24

DETAILS:

    On 12-16-98 at 1530 hours I received information during roll call that Rabbit was wanted for a robbery of the Tall Can Liquor Store at 301 Pacifica Street. The information included a physical description and a color photograph. At 1830 hours I was driving south on Main Street and saw Rabbit standing in line at the XLT Bus Depot.

    I spoke with him and he told me he was Rabbit. I verified the warrant was valid and arrested him. I booked Rabbit at the El Fuego City Jail.

J. Fava #24

Figure 7-3   One-Person Arrest Report Form.

EL FUEGO POLICE DEPARTMENT

ARREST REPORT 99-098418

ARRESTED: 1.   Adams, Paul DOB 11-27-55

CHARGE: 459 PC Burglary

                2.   Barkton, Steven DOB 3-17-46

                      459 PC Burglary

LOCATION: 423 Raindrop Lane, El Fuego

DATE and TIME: 2-17-99 1500 Hours

OFFICERS: M. Conklin #199

                R. Sproul #7654

DETAILS:

On 2-17-99 About 1445 hours we received a radio call of a silent intrusion alarm at 423 Raindrop Lane. We arrived and set up a perimeter and could see Adams and Barkton through the front window. They were standing in front of an entertainment center and were disconnecting wires from the video recorder. Using the car's public address system I announced our presence and ordered them to come out. As they walked out the front door, Sproul arrested them and put them into separate cars.

I spoke to Billy Huntly, 426 Raindrop Lane, (213) 555-1212, and learned that Kathy Czaban (213) 555-1212 is the owner of the house and was away on vacation. I phoned her at her hotel in Los Angeles and learned that Barkton is a former boyfriend who did not have permission to be in the house. There is no loss at this time, but Czaban will make a complete check when she returns.

I booked Barkton and Adams at the El Fuego City Jail.

M. Conklin #199

Figure 7-4   Multiple-Person Arrest Format Report.

ARRESTED: 3.
CHARGE:
LOCATION:
DATE AND TIME:
OFFICER:
DETAILS:

In either case, you would begin the narrative below the heading DETAILS with the date, time, and how you got involved.

Questions about how much detail is needed and what kind of information to include in an arrest report are frequently asked. Ideally, every arrest report would include everything that happened, answer every question anyone reading it might have, and allow for a quick resolution to the matter because of its completeness. Realistically, this is not likely in many cases, but there are some things that are necessary in every case:

1. Identify all people present at the location of the arrest or who participated in it in any way.
2. Identify and explain all injuries to any person involved and how they occurred.
3. Clearly report statements made by the suspect.
4. Clearly describe the probable cause for the arrest.
5. If there are multiple charges, make sure each charge is supported with probable cause.

It is not the arresting officer's obligation or duty to prepare a one-sided, biased, or slanted report. This serves no purpose other than to delay the criminal justice process. It is the arresting officer's responsibility to report all the relevant facts in a complete, clear, concise, and accurate manner that will allow all those who read the report to form their own opinions and draw their own conclusions. The circumstances of how you get involved in cases ending with an arrest are many times out of your control. One thing within your control and that you are able to manage is your ability to write a good report that clearly shows who, what, where, when, why, and how you got involved, establishing a factual basis for what you did. The key to writing a good arrest report in any situation is to know the law, know your agency's policies, and do your job within these parameters.

SUMMARY

Arrest reports should be completed whenever a person is taken into custody and as soon after the arrest as possible. Whether an investigator uses a fill-in-the-blanks and narrative-type report or a straight narrative report, the most important thing to establish in an arrest report is the probable cause used to take the suspect into custody. Other important items that should be included in the report are the names of everyone who was there, details about injuries to anyone present, and any statements the suspect made.

REVIEW

1. An arrest report is needed whenever someone is taken into custody.
2. The single most important thing to be established in an arrest report is the probable cause.
3. The length of an arrest report depends on the facts of the case.
4. Use the rules of narrative writing to complete the narrative of the report.
5. Use short, clear, concise, and concrete words.
6. Limit unsupported opinions.
7. The report must be unbiased.

EXERCISES

1. Using the following fact pattern, prepare an arrest report using today's date and time.

   FACTS: You are a patrol officer for the El Fuego Police Department and are on duty as Unit 6. You are on patrol traveling west on Carson Street. As part of your normal and routine patrol duties, you drive north on Faculty, and as you are approaching the T intersection of Village and Faculty, you see a 1997 red Corvette, driving east on Village, come up to the stop sign at Faculty and make a left turn onto Faculty without stopping at the stop sign. The wheels of the Corvette never stop and actually break traction as the driver accelerates out of the turn. Your best estimate of the car's speed is five miles per hour. The license plate on the Corvette is a personalized plate: SEEYA. You make a car stop at the next cross street north of the intersection, which is 400 yards away at the intersection of Faculty and South Street. When you talk to the driver, he tells you that he usually does not stop at stop signs if there is no traffic because it is inconvenient. He gives you his operator's license and you see that his name is Joseph Paul White and his birth date is July 4, 1963. He also tells you that he never pays his tickets and probably has a warrant out for his arrest. You check his driving record as you write the ticket on Citation number EF9987659, and when you are almost finished, you get the following radio message from the dispatcher.

   Dispatch: Unit 6.

   You: Unit 6, go ahead.

Dispatch: Unit 6, there is an outstanding burglary warrant for Joseph White, DOB 7-4-63. Bail is $15,000 and the warrant number is F4591015.

You: 10-4.

After the driver signs his ticket, you arrest him for the warrant and take him to the El Fuego City Jail where you book him. You put his copy of the traffic ticket in his property. Before you took him to jail, he asked you to lock up his Corvette and leave it where it is parked. The registration in the car showed that the driver is the owner and that he lives at 3007 Raceway Lane, El Fuego.

2. Using the following fact pattern, prepare an arrest report with an evidence report. Use today's date and time.

FACTS: You are a patrol officer for the Big Rock Police Department and are working with your partner Officer Ron Ho as Unit 2. As part of your daily briefings, you are familiar with an ongoing problem of property damage occurring in the industrial area in which unknown suspects have been spray painting graffiti on the buildings. You have had several conversations with the local business owners about the problem, including Mr. Steven Luck who owns the Good Luck Auto Repair Shop. Mr. Luck has told you several times that if you ever catch someone painting on his property, he will cooperate and support prosecution. While you are patrolling an industrial area, you drive around the corner from east bound Container Lane to south bound Cement Avenue and both you and your partner see a man spray painting the letters BG onto the side of the Good Luck Auto Repair Shop. Your partner jumps out of the car and tells the man to stop, but he finishes the painting. As soon as the painting is done, he throws the can down and starts running away, south on Cement. Officer Ho is in excellent condition and easily catches the suspect after a chase of only 50 yards. The suspect gives up without a struggle and Officer Ho handcuffs him. While Officer Ho is walking the suspect back to the patrol car, you picked up the spray can. It is a 10-ounce can of red paint made by Brand X. The suspect tells you his name is Trevor Robers and his birthday is February 14, 1985. You and Officer Ho take the suspect to the Big Rock Jail and book him for vandalism.

3. Review the following report and identify any issues. Using the arrest report format, the facts provided, and the rules of narrative writing, prepare an arrest report. Use today's date and the time shown in the report.

FACTS: Today at approx. 1015 AM I was taking a theft report at the Brand X gas station at Roscoe & Beach. I heard a broadcast go out that a susp. was in a bank across the street & that he was a possible forger.

I left the report and drove across the St. to the bank. Motor Off. Najjar & Brown assisted. I detained susp. Valley until we got a story from the bank. I ran a check on Valley & 1 outstanding warrant came back to his name. His drivers lic. & photo on it looked phony to me, his face didn't match his drivers lic. Valley spoke with a German accent but said he was born and raised in Los Angeles.

I advised Valley of his Miranda rights. He waived to talk to me by saying he'd tell me his story. He said he took a cab today from the airport in Los Angeles enroute to San Diego. He stopped at the Bank of Money in El Fuego to make a withdrawal from his savings account that he opened at another branch of the bank in San Diego last year. He came alone. He said he deposited two checks totaling approx. $7000 in his San Diego acct. 3 weeks ago. The checks were 2nd party checks from a friend of his in New York. Today he wanted to withdraw $2500 cash of that money.

After investigation it was ascertained that Valley (if that is his name) tried to steal $2500 from the Bank of Money. Three other males, probably friends of his split the location of the bank when I first arrived, & were subsequently apprehended. They all had German type accents and said they were from Los Angeles. I arrest the man purporting to be Clifford Valley on the warrants & additional charges relating to the attempt theft at the bank.

ADDITIONAL FACTS: The Bank of Money is located at 10 Dollar Lane, El Fuego, CA 90001, (132) 555-1212. The person you talked to at the bank is Yolanda Sweet, the branch manager. According to Ms. Sweet, Valley and three other men came into the bank and got in the teller line. All had accents but she could not identify them but thought they were German or Russian. Valley presented his driver's license and wanted to withdraw $2500 from an account in San Diego. The account was in the name of C. Valley, but there was a hold on the account. You called the bank in San Diego and spoke to manager I. M. Honest who said that two stolen checks totaling $7000 had been deposited 4 days ago and had been returned to the bank. The warrant you found for Valley is F336514, for Forgery, Bail $150,000, issued on 5-2-2002 in the West Court Judicial District.

4. Review your local newspaper to locate a criminal case in which an arrest was made and the case subsequently filed. Visit the court clerk's office and see if the arrest report is available for review. If so, see if you can identify the probable cause for the defendant's arrest. Does the report follow the rules of narrative writing?

5. Using the following fact pattern, prepare an arrest report using your state and zip code for all address needs.

FACTS: The date is August 21, 2006, at 11:00 A.M. You and Officer Tom Wagner are working together in a marked police car on routine patrol when you get the following radio call, "26X respond to a vehicle tampering in progress in the 3000 block of Elm." You start heading that way and just before you get to the address, a second broadcast is received as follows, "26X the informant, John Brown states the suspect is trying to steal the car, a white Explorer." From a short distance down the street, you observe a white adult male sitting in the driver's seat with his left leg out of the car and his foot on the ground. The hood of the Explorer is partially open. Because you approached from the rear of the victim's car, the suspect did not see you coming. Just as you are pulling up to stop, the suspect looked at you, got out, and took off running. Your partner was able to get out of the police car pretty quickly and order the suspect to stop, which he does. When you walk up to the open car, you can see a remote starter switch lying on the floor with the wires still attached under the hood. You talk to the suspect and ask him what he is doing and he tells you that he was resting in the car. At this point, you and your partner discuss what has happened and, based on your experience and training, believe you have interrupted an auto theft in progress and take the appropriate action. A man approaches you and claims to be the informant and owner of the car, which is a white, 1999 Ford Explorer with California license plate 6H312T. You run the plate and find it is registered to John Brown, 3110 Elm, El Fuego. The victim states that he saw the suspect walking down the street checking car door handles. He states that when the suspect got to his car, he stuck something into the door lock and then opened the door. The victim states that his car was locked when he parked it earlier in the morning around 9:00 A.M. The victim further states that after the suspect got the door open, he unlatched the hood and put some wires under the hood and then got back in the driver's seat and tried to start the car. Right after that, the police showed up and caught him, and the victim wants to prosecute him. At this point, you check the car more thoroughly and see that one of the connectors of the remote starter switch is attached to the battery and the other one to the starter solenoid. You take some pictures of the remote starter switch and tell the suspect he is under arrest for attempted auto theft. At this point, you give him his rights, and he says he knows them better than you do. Before you can ask him any questions, he tells you he is an escapee from the El Fuego County Jail and does not want to say anything else. After you collect the evidence, you and Officer Wagner book the suspect at the El Fuego City Jail.

VICTIM: John Ray Brown DOB 6-10-86
3110 Elm, El Fuego
Home Phone (555) 993-3128
Work Phone (555) 983-7714
Business: Salesman, Computer City,
101 Main, El Fuego
SUSPECT: Frank Johnson DOB 10-8-51 Escapee, no address or phone

QUIZ

1. When a suspect is arrested for multiple charges, only the most serious charge should be established with probable cause.

   a. True
   b. False

2. What is the most important thing to be established in an arrest report?

3. What are two basic types of arrest report formats?

4. What is the format for side headings for a straight narrative arrest report?

5. How should the narrative of an arrest report begin?

6. When should an arrest report be completed?

7. What is the minimum length of an arrest report?

8. Injuries to the suspect should be explained only if they are serious. Is this true or false? Explain.

9. Generally speaking, the arresting officer should include an opinion in an arrest report. Is this true or false? Explain.

10. It is the arresting officer's duty to write a biased report that will ensure a suspect is convicted. Is this true or false? Explain.

8

Writing the Interview

KEY POINTS

The process of investigators talking to people for the purpose of getting information is called *interviewing.* Another term that is sometimes used in this regard is *interrogation,* but the two are different and have different purposes. An **interview** is usually less antagonistic, is more relaxed, and has the general sense that the questioner is drawing information from the person with many questions. It is a facilitative process in which the interviewer helps the interviewee to remember things and explain them in detail. An **interrogation** is much more focused; it has a heightened sense of anxiety and tension. In an interrogation, it is very clear that the person being questioned is the focus of the investigation or close to it. It is not uncommon to interview a person in the initial stages of an investigation and then bring the person back for an interrogation at a later time. This chapter will focus on the three key parts of interviewing as they relate to report writing. The first part is the preparation; the second is the interview itself, and the third is how to put what was learned into a usable written form. Knowing how and when to conduct an interview and report it accurately is key to a successful investigation. There is nothing better than getting solid information from the most reliable source—those involved in the matter. This chapter will provide the background and some helpful hints on how to achieve success in this area.

KEY TERMS

Behavior baseline

Developing rapport

Establish rapport

Interrogation

Interview

Interview worksheet

Invoking rights

Miranda admonishment

Miranda waiver

Nonverbal response

Preparation for the interview

Verbal response

THE PURPOSE OF THE INTERVIEW

When superior investigators begin a case, their one and only goal is to find the truth. Some of the means they use to achieve the goal have been discussed. One of the most often used techniques is to talk to people. This may be the best way of getting new information, verifying information, and finding out about all the minute details and innuendos of the case. Talking to people about the case can give the investigator a general picture of what a victim, witness, or suspect does not know. Once this information is in hand, it needs to be memorialized, which is a fancy way of saying it needs to be written in a report. What percentage of the whole each of these comprises is open to debate, but it is almost guaranteed that if any of the three is lacking, the chances that this part of the investigation will be troubled increases on a sliding scale.

I think the single most important thing an investigator can do to ensure a successful interview is to be prepared. Not all investigators are good interviewers, but almost all good investigators are good interviewers. This part of the job is vital and worthy of as much time as the case allows. There are a number of factors that contribute to the skill set of a successful interviewer including having a general awareness of local, national, and world events; being familiar with the facts of the case; having the ability to relate to people; being able to establish rapport; being a good listener; shaping questions based on the responses of the person being interviewed; and reading the body language of the person being interviewed and recognizing that the nonverbal clues given by body language match what the person is saying. With this nonscientific list of skills, the investigator is ready to prepare for the interview. A rule of thumb for investigators is that they can never know too much about the person they are going to talk to, or too much about the matter they are going to be talking about.

If there is such a thing as a general description of a typical interview, it might be something like this. After a thorough preparation, the investigator conducting the interview begins by asking several questions that are so generic that it is very unlikely the interviewee would lie when answering them. The interviewer starts with these not only to hear what the person says but also to see how he or she acts when doing so. This is the first time the interviewer will have an opportunity to compare the **verbal** and **nonverbal** responses, and it establishes a **behavior baseline**. Verbal responses are the things the person says including the manner in which he or she hesitates, repeats the question before answering, or stammers. Nonverbal responses are the things the person does while answering. Does the person groom himself or herself, rub his or her nose or mouth, shift in the chair, slump or sit upright while answering or immediately thereafter? These clues give the investigator a picture of what the interviewee looks like when telling the truth. As the questions become more focused, the person may react differently when saying something that is not true. An investigator who sees and hears a conflict between verbal and nonverbal answers can explore this anomaly more thoroughly; it could become useful as the totality of the information and evidence is reviewed. When satisfied that all the information has been obtained, the investigator reviews it with the interviewee who should agree that it is correct. The person is asked to confirm that all information he or she has given is truthful and that the investigator made no promises or threats to obtain the answers. The interview concludes with a request for the interviewee to give a written statement. If one is given, it is signed by the person being interviewed and booked as part of the case evidence. The interview should be considered successful if information useful to the investigation was gained.

PREPARING FOR THE INTERVIEW

Preparation for an interview should begin with the goal to learn as much background information as possible about the person to be interviewed and about the possible crime. Knowing all that an investigator can helps him or her develop the questions to use to make the initial "read" of the person and how he or she acts when giving presumably truthful answers. Again, this is where the investigator establishes the person's behavior baseline. Many investigators prepare a questionnaire to help structure their interviews. One example is the **interview worksheet** (Figure 8-1). The information in this sheet is by no means comprehensive but is a starting point for each investigator to develop a workable format of his or her own design. One of the final steps in the preparation is to put together the information the investigator knows about the person to be interviewed. The value in having such a worksheet completed is that it allows the investigator to form baseline questions, can give a sense of what the subject is like, and can provide some leads to follow in **developing rapport** with the person. The investigator may not be able to find out everything on the prepared worksheet, but whatever he or she can verify during the interview could be helpful.

The reasons why people admit they did something illegal or improper can be as varied as the people themselves. Some feel a relief from the guilt their actions caused, others might see their

INTERVIEWEE:

BIOGRAPHICAL

1. Name	Nickname
2. Date of birth	Place of birth
3. Title	
4. Home address	
Phone _____	
5. Vehicles	
6. Business address	Occupation
Phone _____	
7. Military service	

EDUCATIONAL

8. School
9. Sports/Achievements

FAMILY

10. Marital status	☐ Single	☐ Married
11. Children ☐ Yes ☐ No		

BACKGROUND

12. Criminal record   ☐ Yes   ☐ No   If yes, describe: _____
13. Hobbies
14. Accomplishments
15. Tragedies in life

ASSOCIATES

16. Friends	Family
17. Home	
Biz addresses	
Phone _____	
18. Criminal records ☐ Yes ☐ No If yes, describe: _____	

LIFESTYLE

19. Medical history   _____
20. Favorite hangouts   _____
21. Preferred style of dress   _____
22. Known political views   _____
23. What upsets the person   _____

Figure 8-1   Interview Worksheet.

admission as an effort to begin a new part of their lives based on honesty, and still others may admit involvement because they are tired of living a lie. A fairly accurate rule of thumb, however, is that people are not going to admit their misconduct to people they do not like, so developing some rapport with them is helpful. This does not mean that investigators should ingratiate themselves with subjects of an investigation or do anything illegal or improper to gain their trust and help, but there is nothing wrong with being considerate, polite, and professional and showing an interest in the person the investigator is going to talk to—a person who may be able to help the investigator find the truth and successfully close a case.

CONDUCTING THE INTERVIEW

If the first step in the interview process is preparation, the second step would be selecting the time and location for the interview. Time and circumstances will influence this part of the investigation, but the bottom line is that the investigator needs to select the time and place. The goal here is to pick a time and place that will be a quiet and private place that is free of distractions. Investigators should avoid starting an interview and then having to stop and reschedule or relocate it, somewhere else. Whether the investigation is administrative or criminal, some external factors can influence the decision in this area. A generally accepted premise when selecting the time and place of administrative interviews is that the interview should take place at the subject's workplace and during his or her regularly scheduled work hours. Another major consideration is that many employees are entitled to representation during the interview, so advance notice must be given to allow the employee's representative to be found and scheduled. If the investigator is conducting a criminal investigation and the circumstances dictate, an interviewee may also be entitled to legal representation, and if this request is made, it must be honored. The presence of another person in the room during the interview should not be a major concern for the investigator with one exception. If the person has a representative, the investigator conducting the interview should have a partner to take notes and act as a witness.

Circumstances may also require the investigator to give a suspect a **Miranda admonishment** prior to asking questions related to the crime. Generally speaking, a successful Miranda admonishment can be given and documented as follows. First, the investigator should have a printed Miranda admonishment that includes the two waiver questions. Second, the investigator should read the admonishment and questions verbatim to the suspect. Admonishment is part of the interview process, and the investigator needs to have clear and positive responses to the two waiver questions before continuing. Anything less than positive responses should be considered the interviewee's refusal to talk to the investigator. An interviewee who tells the investigator that she or he does not understand her or his rights or does not want to talk is **invoking** her or his **rights**. This means that the questioning must stop, and the investigator must move on to other aspects of the investigation. The interviewee's positive and convincing answers that she or he understands her or his rights and wants to talk without a lawyer present are called a **Miranda waiver**. Third, the investigator should quote the interviewee's responses in the written notes.

Assuming that the investigator receives a waiver and the person agrees to talk, the investigator should start asking questions. Learning how to do this proficiently and successfully can take years, but some basic things will help the investigator develop good habits. First, the investigator writes or "scripts out" the first 20 to 30 questions to ask and is prepared to note the person's answers. Being able to write fast and write the entire answer is great, but it is not necessary at this point. Structure the questions so that they require more than a yes/no answer because you want to get the person in the habit of talking and explaining things. Second, the investigator should ask one question at a time and then be quiet and let the person answer it. This is difficult for many new investigators, but it is critical to the process. The person should have a chance to answer without interruption or badgering. Third, if the suspect gives short, curt answers, try to get more explanation by asking questions such as, "What do you mean?" "What happened then?" The idea is to let the person explain what happened. Fourth, ask a person who is open and talkative to give a short version of what he or she said, did, or heard. As discussed in an earlier chapter, the Thirty word version of the story will help you to get a quick read on what the interviewee might be able to report when the investigator delves more deeply. Fifth, if the person is willing to tell what happened, let him or her talk without interruption. Once the person has finished the story or starts to veer off course, the investigator can review the facts with the person. Sixth, once the investigator has gotten what appears to be the whole story, review it with the interviewee for accuracy. The investigator can say something like, "Joe, I want to review with you what we talked about here today. What I heard you say is . . . ." Then go over what he said and give him a chance to correct anything that is incorrect. The goal of doing this is to give the person a chance to learn what the investigator heard him say so that later on when it is repeated in court, there is no confusion and the person cannot deny the answers given. Seventh, once the investigator has reviewed the person's story, ask the person about his or her treatment during the interview, whether he or she had a chance to answer what was asked, and whether the person had told the truth about everything. Eighth, after reviewing what the interviewee said and he or she agrees that it is accurate, the investigator can ask the person to give a written statement. A pad of paper and a pen should be available and given to the person. The investigator should never tell the person what to write, but if he or she asks what to write, the investigator can say something like, "Write what you told me." Some people will say they don't know how to write and ask the investigator to write for them. If this is the case, I feel it is acceptable for the investigator to write what the person dictates as long as a witness is present. This will prevent the person from saying later that the investigator made the responses up. When a written statement is completed, the investigator asks the person to sign it and initials it. It becomes part of the evidence in the case.

The goal of all this work is not to get a confession but to get the truth in the form of information received in a legal and professional manner. Interviewing people is probably the best way to find out what they know, did, or saw. It can be a very efficient way to move forward in a case, but getting the information is only part of the task. While being a good interviewer is a plus, being able to write down what the investigator learned in the investigative report is just as important.

WRITING THE INTERVIEW

Many young investigators believe it is important to write down word for word what was said in an interview. It is not uncommon to see a report with the question followed by an answer, question–answer, question–answer, and so on. While accurate, this kind of writing is overkill. I suggest a less rigid and time-consuming style that provides the essence or gist of the information gained while meeting the needs of the investigation. When using this method, the investigator does not need to write each question and each answer. Rather, an introductory sentence followed by what was said in general terms is usually sufficient. For example,

> I interviewed Smith at his home and he told me. . . .
> I read Thompson his rights per Miranda, and in response to the two waiver questions, he said "yes" and "yes" respectively. Thompson then told me. . . .

There is generally no need to write the questions asked or quote what the suspect said in an interview with three exceptions. The first regards Miranda. The investigator should always quote what the suspect said when she or he acknowledges the rights and waives them. The second instance is when a suspect admits guilt. His or her exact words are very helpful in prosecuting the case and should be quoted if at all possible. The third instance in which a quote is needed is when a crime victim reports the words the suspect said during the commission of the crime. These words are often part of the suspect's Modus Operandi and can be helpful in connecting a suspect to a string of crimes.

SUMMARY

The art of interviewing is a skill that can be learned and used successfully by almost anyone. The key to obtaining an admission of guilt involves no tricks or clever word choices but performing only good solid investigative work in which the investigator uses the law and circumstances to his or her advantage. The investigator should be prepared and pay attention to the nonverbal clues and verbal responses the interviewee provides. The questions should be direct, and the person should have enough time to answer before another is asked. Finally, the investigator should review the information gained during the interview and write an accurate report of it. As stated at the beginning of the chapter, the preparation usually facilitates a successful interview.

REVIEW

1. Preparation is key in being a successful interviewer.
2. Build rapport with the person to be interviewed.
3. Select a time and place for the interview to your advantage.
4. If applicable, read the Miranda admonishment and quote the response.
5. Ask direct, open-ended questions and listen to the answers.
6. Give the person time to answer.
7. Review with the interviewee what he or she told you for accuracy.
8. Write the gist of what the suspect said.

EXERCISES

1. Practice reading the Miranda admonishment and writing about it. Be sure to quote the waiver. Will you be able to accurately testify about the interviewee's response?
2. Role-play an interview with a classmate. Write the results of the interview and have your partner check it for accuracy.
3. Watch a news program that includes an interview and write the results in a report format. Were you able to write an accurate report?

QUIZ

1. What is the most important part of the interview process?

2. What is the purpose of an interview?

3. If a Miranda admonishment is needed, what is the generally accepted method of giving it?

4. When you interview a suspect, you are trying to get a confession.
   a. True
   b. False

5. Is it always necessary to quote a suspect's answers?

6. What is the purpose of scripted questions for an interview?

7. What is the purpose of the interview worksheet?

8. What is a common mistake most investigators make in the interview?

9. Why is rapport with a suspect important for an investigator to establish?

10. What are the three parts of the interview process?

9

Writing Search Warrants

KEY POINTS

Whether you work in a small or large agency, if you investigate enough cases, you will someday need a search warrant. In this chapter, we start with a discussion about the key parts of a search warrant and the process of obtaining one. Next we cover the key part of the writing process in which you establish your qualifications as the affiant. We conclude with a review of the use of the rules of narrative writing to complete the warrant, how to incorporate existing reports into the affidavit, and how to list the evidence seized in the return to the warrant.

KEY TERMS

Affiant

Affidavit

Evidence list

Exhibits

Probable cause

Reasonable particularity

Return to the warrant

Search warrant

Statement of probable cause

At some point in your investigative career, you will encounter a situation where you need some information that is not readily available. Your need might be for business records or access to a crime scene that requires more than consent—in short, you will need a search warrant. When this situation arises, you will need to have a working knowledge of search warrants and the process needed to obtain one. The purpose of this chapter is not to provide an all-encompassing report on this investigative tool but rather to establish a writing strategy for completing the necessary documents to obtain, serve, and return a search warrant.

SEARCH WARRANTS

One of the assets a good investigator must have is a working knowledge of search warrants. In today's world, a lot of valuable information can be obtained, but it is more and more likely that the person or business in possession of what you need will not voluntarily part with it. This makes it essential that you know what to do and how to do it if and when you face this challenge. This section is not an attempt to cover every aspect of search and seizure as it relates to search warrants or to qualify you as an expert in the area of search warrants. Rather it is to give you a basic overview of what a search warrant is, how to write one, and what the process is to obtain, serve, and return one.

What is a search warrant? Although exact definitions may vary from jurisdiction to jurisdiction, the thirty word version is that a **search warrant** is an order from a magistrate to a peace officer to go to a particular place, search for a particular thing, and if it is found, bring the thing back and show it to the magistrate who issued the warrant. A search warrant can be issued only when probable cause exists, and your job as an investigative report writer is to put the probable cause into the proper format. Generally, the **probable cause** must show that one of the following things has recently happened or is occurring:

1. Property is stolen or embezzled.
2. Property or things were used as the means of committing a felony.
3. Property or things are in the possession of any person with the intent to use it as a means of committing a public offense or in the possession of another to whom he or she may have delivered it for the purpose of concealing it or preventing its being discovered.
4. Property or things to be seized consist of any item or constitute any evidence that tends to show a felony has been committed or tends to show that a particular person has committed a felony.
5. Property or things to be seized consist of evidence that tends to show that a child has been or is being sexually exploited.
6. There is a warrant to arrest a person.

A search warrant consists of three distinct parts: the search warrant, the affidavit in support of the search warrant or statement of probable cause, and the return to the search warrant.

THE WARRANT PROCESS

When the need for a search warrant exists, the **affiant**, who is the person that prepares the warrant, compiles all of the facts and evidence into written form, properly completes the search warrant forms, and presents the search warrant and affidavit to a deputy district attorney for review. Once the deputy district attorney agrees that the information in the warrant and affidavit support its issuance, the affiant takes the warrant to a magistrate. The magistrate swears the affiant in, much as a witness is sworn in before testifying, and reviews the warrant and affidavit. If the magistrate agrees that the warrant and affidavit have established probable cause for a search, the affiant signs the document and then the magistrate signs. At this point, the warrant must be served and returned to the court within ten days.

After serving the warrant and seizing the property listed in the warrant, the affiant prepares the return to the warrant. The return is a written list of all evidence seized pursuant to the service of the warrant. Once the return is prepared, the affiant takes the original search warrant, the affidavit, and the return back to the magistrate and shows the magistrate what was seized. Both the affiant and magistrate sign the return, and the affiant then gives all three parts of the warrant to the court clerk, who files it. Generally speaking, at this point all the information contained in the search warrant, affidavit, and return become public record and anyone may obtain a copy of them. (See Figure 9-1.)

HOW TO WRITE A WARRANT

The search warrant is the part of the document that contains the description of the place to be searched and a description of the evidence you are looking for.

Describing the Premises

The description of the premises to be searched must be so complete that any peace officer could pick up the warrant, read the address, and be able to find the location to be searched. This can be accomplished by describing the premises with the address, city, county, and state, followed by a physical description of the property, including style of construction, color, where the address numbers are attached to the building, and the geographical location of the structure in relation to fixed reference points. For example:

123 Elm, City of Long Wave, County of Los Angeles, State of California, a single-story wood and stucco structure with a

STATE OF CALIFORNIA - COUNTY OF ORANGE   SW NO. _____

SEARCH WARRANT AND AFFIDAVIT
(AFFIDAVIT)

_____ declares under penalty of perjury that the facts expressed by him/her in
(Name of Affiant)
the attached and incorporated **Statement of Probable Cause** are true and that based thereon he/she has probable cause
to believe and does believe that the articles, property,and persons described below are lawfully seizable pursuant to Penal
Code Section 1524, as indicated below, and are now located at the locations set forth below. Wherefore, affiant requests
that this Search Warrant be issued.

_____, **NIGHT SEARCH REQUESTED:**   YES [   ]   NO [   ]
(Signature of Affiant)

(SEARCH WARRANT)

THE PEOPLE OF THE STATE OF CALIFORNIA TO ANY SHERIFF, POLICEMAN OR PEACE OFFICER IN THE COUNTY OF ORANGE:
proof by affidavit, under penalty of perjury, having been made before me by_____
—
(Name of Affiant)
that there is probable cause to believe that the property or person described herein may be found at the locations set forth
herein and that it is lawfully seizable pursuant to Penal Code Section 1524 as indicated below by "x"(s) in that it:

_____ was stolen or embezzled.

_____ was used as the means of committing a felony.

_____ is possessed by a person with the intent to use it as means of committing a public offense or is possessed by
     another to whom he or she may have delivered it for the purpose of concealing it or preventing its discovery.

_____ tends to show that a felony has been committed or that a particular person has committed a felony.

_____ tends to show that sexual exploitation of a child, in violation of Penal Code Section 311.3, or possession of matter
     depicting sexual conduct of a person under the age of 18 years, in violation of Section 311.11, has occurred or is
     occurring.

_____ there is a warrant to arrest the person.

YOU ARE THEREFORE COMMANDED TO SEARCH: (Premises, vehicles, persons)

FOR THE FOLLOWING PROPERTY OR PERSONS:

AND TO SEIZE IT/ THEM IF FOUND and bring it/ them forthwith before me, or this court, at the courthouse of this court. This **Search
Warrant** and **Affidavit** and attached and incorporated **Statement of Probable Cause** were sworn to as true under penalty of perjury
and subscribed before me on (date) _____ , at _____ A.M. / P.M. Wherefore, I find probable
cause for the issuance of this Search Warrant and do issue it.

KNOCK-NOTICE EXCUSED:☐

Figure 9-1 Search Warrant and Affidavit. (As described in California Penal Code)

brown shake roof. The front door of the residence faces south, and the number 123 is attached to the fascia directly over the front door. There is an asphalt driveway along the east side of the residence, and the house is the fourth structure west of Vine Avenue on the north side of the street.

Describing an Open Field

If the premises or location to be searched is an open field, the description might be:

An undeveloped lot on the southwest corner of Beach Boulevard and Main Street, City of Long Beach, County of San Diego, State of California. The lot is bordered on the west by a 6-foot-tall block wall, on the north by the south curb line of Main Street, on the east by the west curb line of Beach Boulevard, and on the south by the north wall of the Acme Toy Company. The entire lot is visible from Main Street and Beach Boulevard and has no structures or trees on it. Although there is no known address for the lot, it is in the 8000 block of Main Street and the 200 block of Beach Boulevard.

Describing a Remote Area

In some cases, searches must take place in remote areas far from developed streets and highways. In these cases, the description might include a picture or a reference to a map to help find the exact location. For example:

An abandoned wooden shack one-half mile north of Fire Road 6 and 2 miles west of Highway 14 in the Southern National Forest, County of Los Angeles, State of California. The area is accessible by air and four-wheel-drive vehicles and appears to be without utility connections. The area of the premises to be searched is shown on page 1134 of the 1997 edition of the Maps-R-Us Road Atlas at coordinates KT-89. A copy of this page has been marked with an X to show the location of the shack and is attached herein and incorporated as Exhibit A. A photograph of the shack has been attached herein and incorporated as Exhibit B. The shack is a single-story, wooden structure with no windows and a door on the west side.

Describing Vehicles

If your investigation requires that you search a vehicle, your description might be:

A 1946 Chevrolet panel truck, dark gray in color with California license plate 1G63410. The vehicle is missing the left front fender and the right rear wheel. It is parked on Elm Street, south of Orange, in the City of Chico, County of Butte, State of California. The car is sitting on concrete blocks, with the wheels off the ground. The vehicle was last registered in 1964, and there is no current registered owner information available per the Department of Motor Vehicles.

Business Records

Obtaining business records such as telephone subscriber information or bank account information is likely to be part of any long-running investigation and should be looked at the same way as any other request for information. One difference in serving a search warrant for business records is that you are usually not going to actually conduct a search of the business. Normally, you will notify the custodian of records ahead of time and let the custodian know what records you are looking for. Your search warrant is then presented to the proper authority, and the records are turned over to you. If you need to access a safe-deposit box at a local bank, your description of the premises to be searched might be:

Safe-deposit box 1225 located in the vault of the Bank of Money, 121 Dollar Drive, City of Cash, County of Coin, State of California. The Bank of Money is in the Willow Branch Shopping Center and sits on the northeast corner of Dollar Drive and Walnut. The Bank of Money is between the Good Burger Restaurant and the Lifetime Appliance store. The bank is a three-story masonry building, brown in color, with the number 121 attached to the east side of the building directly over the door.

Telephone Records

If you were looking for subscriber information at a phone company, perhaps the description would look this way:

West Coast Telephone Company, 6426 Cellular Drive, City of Big Tree, County of Oakland, State of California. The requested records are kept by the custodian of records in Room 100 on the first floor. The building is on the south side of Cellular Drive between 64th and 65th Streets. I do not intend to conduct a search of the premises but rather to turn the search warrant over to the custodian of records and receive the requested information.

Remember that the test of whether your description of the premises to be searched is good enough is whether any police officer could read your description and then find the location or premises to be searched.

Describing What You Are Looking For

The description of the property you are looking for must meet the standard of **reasonable particularity**, which means that it is good enough that the average person could read the description and pick the item out of a group of similar objects. There are thousands of items that can be the subject of a search warrant, and this book is not designed to provide examples of them. The rule of thumb here is that the property you are looking for must be described with reasonable particularity, and the need for the items you are seeking must be supported by probable cause in the affidavit. Local prosecutors may have preferred formats or descriptions of the more common items sought such as phone records, bank account information, narcotics, and bookmaking records. It is always advisable to seek their input to ensure that your warrant is complete and sufficient.

WRITING THE AFFIDAVIT

The **affidavit**, or **statement of probable cause**, is the part of the document that contains all of the facts, information, and evidence that establishes the probable cause for the issuance of the warrant. In addition to the facts of the case, the affidavit also includes a description of the affiant as a way of introduction to the magistrate. (See Figure 9-2.)

The first part of the affidavit should be your introduction. Although there is no ironclad formula for the opening, it should include your job and title, how long you have been in the assign-ment, prior experience, education, and training. It should also include your practical experience in the particular type of case that is being investigated and if and how many times you have testified as an expert in court regarding similar types of cases. All of this helps establish your expertise in the eyes of the judge who is reading your warrant. Beginning the first paragraph with your name is not necessary because it is understood that the person making the application for the warrant is the affiant. Some examples of these introductions include:

Detective

The introduction for a police detective could be in this form:

> I am a member of the Big City Police Department and have been so employed for the past six years. For the past two years I have held the position of Detective and have been assigned to the Burglary Unit. I am a graduate of the Big City Police Academy, where I received instruction in investigating the crime of burglary. I hold an associate of arts degree in administration of justice from Long Beach City College and have attended a 40-hour Basic Investigation class sponsored by the California Department of Justice. I have investigated more than 200 burglaries. I have arrested more than 20 suspects for burglary and receiving stolen property and have spoken with them about their crimes and criminal habits. Based on my experience and training, I am familiar with the types of burglaries being committed in this area and the common motives of burglars.

_____ ,   NIGHT SEARCH APPROVED:☐
(Signature of Magistrate)
Judge of the **Superior / Municipal** Court, _____ Judicial District, Dept. / Div. ____

F026-   (Rev.1/98) Pg.1   f:\forms\sw-af1.198
STATE OF CALIFORNIA - COUNTY OF ORANGE   SW NO. _____
ATTACHED AND INCORPORATED

STATEMENT OF PROBABLE CAUSE

Affiant declares under penalty of perjury that the following facts are true and that there is probable cause to believe, and affiant does believe, that the designated articles, property, and persons are now in the described locations, including all rooms, buildings, and structures used in connection with the premises and buildings adjoining them, the vehicles and the persons:

Figure 9-2 Statement of Probable Cause. (As described in California Penal Code)

Narcotics Officer

An example of someone who is assigned to a narcotics unit might have an introduction something like this:

> I have been employed by the Oak Tree Police Department for five years and have been assigned to the Special Investigations Unit for the past three years. During the first three months of this assignment, I attended an 80-hour Narcotics Investigation class sponsored by the Drug Enforcement Administration and a 40-hour Narcotics Recognition class sponsored by the Federal Bureau of Investigation. I have had contact with more than 200 narcotics users in the Oak County area and discussed with them the use and sales of heroin, cocaine, and other controlled substances. I have had personal contact with more than 50 persons who use and sell heroin in the Oak County area. I have received field training in the field of controlled substances from senior officers Lieutenant D. Roman, Porter Police Department; Sergeant J. Pounds, Central Police Department; Officer G. Najjar, Winston Police Department; and Officer R. Ho, Pico Police Department. I have participated in more than 150 narcotics investigations, have seized cocaine on more than 200 occasions, and have testified as an expert in the field of narcotics in the Municipal Court of Oak County on 13 occasions. Because of my experience and training, I am familiar with the trafficking of narcotics in the southwest portion of the United States, including the methods that users employ to acquire drugs and the methods used by sellers of controlled substances to distribute them.

Vice Officer

Someone working in a vice unit might have an introduction such as:

> I am a police officer in and for the city of Long Branch and have been so employed for the past 13 years. I currently hold the rank of Sergeant, and I am the supervisor in the Special Crimes Unit, with the responsibility of supervising all vice and special assignment investigations. During my career I have worked as a patrol officer, narcotics investigator, patrol sergeant, and watch commander. I have worked as an investigator for a total of six years and have investigated over 2000 cases. I have worked in undercover and administrative assignments, purchased illegal drugs, and participated in hundreds of arrests for violations of narcotics and vice offenses. I hold a bachelor of science in criminology from the University of California, and I am a graduate of the Oak County Peace Officers Academy. I have attended more than 30 specialized training classes and schools sponsored by the

California Department of Justice, California Narcotics Officers Association, the Federal Bureau of Investigation, and the Drug Enforcement Administration. I have worked hundreds of cases with other experienced investigators, and I am familiar with the law as it relates to conspiracy, narcotics, and vice offenses. I have testified as an expert in the Municipal and Superior Courts of Oak County on at least ten occasions.

Although it is helpful to have a strong academic and professional background including recognized certifications, they are not necessary for a person to be an affiant and prepare a search warrant. Everyone has to start somewhere and sometime, and every affiant's introduction will be different. What is important is to have a logical, understandable, and well-written introduction for your warrant. The first step in preparing your introduction is to put together a training record of all the classes and training sessions you have participated in, along with copies of the certificates you received. This is vital, because you may be questioned in court as to the validity of your claims, and having the proof in hand is necessary. It also gives you a way to review the extent of your training and attend classes where necessary to strengthen any areas that are lacking. Once you have assembled the training record, start writing. It may take a couple of attempts to get an introduction that both sounds good and accurately reflects your personal situation. The rule of thumb here is not to exaggerate or overstate your qualifications. Understate your experience rather than go overboard.

Once you have introduced yourself to the magistrate, you complete the affidavit by presenting the facts, information, and evidence about your investigation. This can be accomplished by using the rules of narrative writing and starting this part of the affidavit with the date, time, and how you got involved in the investigation. The rest of the affidavit will explain what happened and how you gathered the evidence you have. Your story must establish the probable cause for believing the facts are true.

If you have occasion to incorporate other reports, documents, or photographs into your affidavit, you may do so by introducing the item in your story and then incorporating it into your warrant by reference. The referenced documents are known as **exhibits** and can be invaluable in establishing probable cause. For example:

> During my investigation, I learned that a similar crime occurred in the adjacent city of Rosarito and that suspects matching the description of the suspects in this case were involved in that crime. A report by Rosarito Police Department Detective John Brown, consisting of 11 pages and documented under case number 06-30481, is attached herein and incorporated as Exhibit A.

By referencing this report, the information in it becomes part of your affidavit. All that remains to be done for this type of inclusion is to mark each page of the attached exhibit with the appropriate label. For example:

Exhibit A Page 1
Exhibit A Page 2

INCLUDING EXPERT OPINIONS

If you include the opinion of another expert in your affidavit to help establish the probable cause, you will need to introduce this person to the magistrate just as you did yourself. Such introduction might look like this:

I reviewed the facts of this case with Billings Police Department Investigator R. D. Moran. Investigator Moran is a police officer for the city of Billings and has been so employed for the past 23 years . . . .

Once you finish establishing your expert's qualifications, you will want to include the person's opinion as to why he or she thinks, the facts of your case establish probable cause.

When you have completed the story of your investigation, the affidavit is almost complete. All that remains is for you, the affiant, to draw some sort of conclusion that the property you are looking for is inside the premises you want to search. This is usually done by saying so in a short paragraph. The wording of the closing paragraph will vary depending on the type of crime and how solid your facts are. There is no minimum or maximum length to the affidavit. The test here is whether the information in the affidavit establishes probable cause to search. If it does, the warrant is ready to present to a magistrate for review.

THE RETURN TO THE WARRANT

The **return to the warrant** consists of a fill-in-the-blanks type of face sheet and a listing of the property and evidence seized during the search. A technique used by many investigators is to list the evidence and property on their agency's property report and attach this **evidence list** to the return. It might be referenced in the return as:

Refer to the attached property report, case number 07-63142, which is attached as Exhibit A.

This written inventory of the seized property is taken back to the magistrate who issued the warrant, along with the original warrant and affidavit. (See Figure 9-3.)

STATE OF CALIFORNIA - COUNTY OF ORANGE
RETURN TO SEARCH WARRANT

_____, being sworn, says that he/she conducted a search pursuant to
<small>(Name of Affiant)</small>
the below described search warrant:

Issuing Magistrate: _____,

Magistrate's Court: **Superior/Municipal** Court, _____, Judicial District

Date of Issuance: _____,

Date of Service: _____,

and searched the following location(s), vehicle(s), and person(s):

R-1

Figure 9-3   Return to Search Warrant. (As described in California Penal Code)

Items attached and incorporated by Reference: YES [   ] NO   [   ]
I certify (declare) under penalty of perjury that the foregoing is true and correct.

Executed at _____ , California   _____
_____
                                                                              Signature of Affiant
Date:_____ at _____ [A.M.] [P.M.]

Reviewed by : _____ Date: _____ at _____ [A.M.]
[P.M.]
                    (Signature of Deputy District Attorney)

F026-   (Rev.7/97) Pg.3   Page ___ of ___

Figure 9-3   *(Continued)*

If your agency does not have a format for this purpose, consider using the one described in chapter 4 and list the evidence seized by category. As discussed, the categories could be:

GUNS

MONEY

DRUGS

ITEMS WITH SERIAL NUMBERS

ITEMS WITHOUT SERIAL NUMBERS

Within each of these categories, list all of the items that belong in it. Let's say you served a search warrant and found two guns, $3,684 in cash, a stolen television, and a kilogram of marijuana in a large plastic bag. Your evidence report, including a chain of custody, might look like this:

1. (1) Smith & Wesson model 19, .357-caliber revolver, blue finish with black rubber grips. Serial number G61141. Found under bed in northwest bedroom by Smith.
2. (1) Walther PPK, .380-caliber, brown finish with brown plastic grips. Serial number unreadable. Found under west end of couch in living room by Wilson.
3. $3,684 in U.S. currency, consisting of (30) hundred-, (12) fifty-, (8) ten-, and (4) one-dollar bills. Found in toilet water tank by Searle.
4. (1) clear plastic bag, tied at the top with gray duct tape, containing marijuana. 1026 grams gross weight. Found inside the kitchen sink cabinet by Najjar.
5. (1) Sony 27-inch color television, black case, serial number TZ641761193G. Found in the garage along the west wall by Ho.

I collected all of the items from the finders and kept them until I marked, packaged, tagged, and booked them into the Big Tree Police Department evidence room.

The main thing to remember is that the description of the evidence and property must be so good that the average person could read your description and pick the item out from several similar items. You must also be sure to establish a chain of custody in the property or evidence report.

The preparation, service, and return of search warrants is not something you are likely to do every day, but you should be prepared to do so at any time. Because procedures can vary from county to county, you should always seek the advice of the district attorney's office when preparing a search warrant and follow the local rules of returning and filing the warrant.

SUMMARY

A search warrant is a very valuable tool for investigators and actually consists of three distinct parts—the search warrant, the affidavit in support of the search warrant, and the return to the warrant. Writing a search warrant will give an investigator the chance to use all of his or her writing skills.

The search warrant is the part that contains the location to be searched and what it is the investigator is searching for. The affidavit introduces the affiant to the judge and contains the story or narrative that explains to the judge why the investigator believes the evidence being sought is in the location to be searched. The return to the warrant contains the list of evidence found during the search. The descriptions of the locations to be searched and the sought-after evidence must all be written with reasonable particularity. The narrative of the affidavit can be written using the rules of narrative writing.

REVIEW

1. A search warrant consists of a fill-in-the-blanks portion and a narrative.
2. The rules of narrative writing should be followed in completing the affidavit or statement of probable cause.
3. Use the first person when introducing yourself to the judge.
4. The description of the premises to be searched must be so specific that any police officer can find the location.
5. The property you are looking for must be described with reasonable particularity.
6. Use the evidence report format to list the items seized.

EXERCISES

1. Using the standard of reasonable particularity, select a vehicle and describe it as if it would be included in a search warrant.
2. Locate the school library, and describe it as if it were a premise to be searched.
3. Write an introduction of yourself as if it would be used in a search warrant affidavit.
4. Using the following fact pattern, prepare a search warrant for telephone records. Use the introduction developed in exercise 3 as a starting point for your affiant introduction. Use today's date and time as a reference, and assume that El Fuego is in the county you live in.

Two weeks ago, you were hired as a special investigator by the California Bureau of Investigations.

For the past week, you have been assigned to investigate all criminal violations occurring at the Do Drop In Bar, 1631 Lager Road, El Fuego.

Yesterday, while you were undercover in the bar, you saw and heard one of the regulars named Billy Biglunch talking about sports and betting on the horses with the owner, Steve Sober. Billy Biglunch told Steve Sober that he had an inside source at the track who gave him very good information that a quarter horse named I'm Not Glue was on and would win the fifth race easily. Based on that, he wanted to bet $100 on the horse to win.

The phone in the bar is on the wall at the east end of the bar. You were sitting on a stool about 3 feet from it. As soon as Biglunch and Sober finished their conversation, Sober used the phone to call in a bet. You saw him dial 1-213-555-9361 and then say, "This is Sober. I want $100 on I'm Not Glue in the fifth." He hung up the phone and told Biglunch, "Okay, your bet is in through me; that was my bookie and we're all set."

You left the bar and, after returning to your office, found that phone service for (213) 555-9361 is provided by Pacific Telephone Company, 1010 N. Wilshire Blvd., Room #621, Los Angeles, California, and that the number is nonpublished.

QUIZ

1. How should you describe the location to be searched?

2. What are the three main parts of a search warrant?

3. What two things are generally contained in the search warrant?

4. What is the test for describing the property you are looking for?

5. What is another name for the affidavit?

6. What should be included in the beginning of the affidavit?

7. What five things should be included in the affiant's introduction?

8. How long should an affidavit be?

9. Which part of the warrant contains the list of the evidence seized?

10. What is the suggested order for listing seized evidence?

10

Issues in Writing

The purpose of this book is to establish a foundation for investigating crimes and documenting the results of these efforts, but experience shows that there are bound to be some questions and some issues that are worthy of further consideration and discussion. This chapter addresses some of these key points and begins with a look at the use of opinions in reports and how the use of opinions can establish probable cause in some cases. Miranda problems and problems with field show ups are also identified, and suggestions are given that will help ease common problems with their use. The chapter concludes with a discussion of how to identify issues in report writing, how to self-correct them, and how to deal with the phenomenon of different people having different uses for the same report.

Expert opinion

Facts

Field identifications

Miranda admonishment

Opinion

Reader use conflict

Report writing issues

Show ups

Several years ago, a fictional investigator became famous for his ability to question people and quickly get to the bottom of the matter at hand. Whenever the person he was questioning would stray off course with an answer, this no-nonsense investigator would tell the person, "Just the facts," and the interviewee would soon after provide a key piece of information needed to complete the investigation. Although this is unlikely in real life, there is a lot of truth to the expression "just the facts" as it relates to investigations.

FACT OR OPINION?

For the most part, investigations need to contain facts. **Facts** are the tangible things we use to make solid decisions, they are things that can be proven, and they are the nuts and bolts of a quality investigation. If facts are at one end of the spectrum, the other end contains things such as suppositions, hunches, gut feelings, and opinions. There is a great deal of difference between a fact and an opinion.

A **fact** is something that can be proven. An **opinion** is a personal belief or judgment that cannot be proven. Although an opinion may be a belief or judgment shared by many, it is not something that can be used to prove certainty. A general rule of thumb is that investigative reports should contain facts only. Opinions should not generally be included; however, there are some occasions when opinions are necessary in an investigative report.

WHEN OPINIONS ARE APPROPRIATE

Many experienced investigators have the mistaken belief that the things they have learned through years of experience and their own common sense are facts that should be included in an investigative report. This occurs when an investigator describes how a suspect accomplished a crime even though there is no available evidence. An example is an auto burglary accomplished by the use of the car key to make entry. In this scenario, the driver parks, hides the key somewhere on the car, and leaves the car unattended for several hours or days. When the owner returns and finds the car unlocked and the car stereo missing, he or she calls the police, and the officer who has been assigned to investigate this crime begins a search of the area. The investigator finds that the hidden key is still in place, there is no sign of forced entry, and there are no witnesses who can tell him or her how the suspect got into the car. Based on this, the investigator decides that the suspect used a shimmying type of tool or some other type of lock pick to unlock the door and in his or her report writes:

The suspect gained entry by shimmying the door lock.

Based on the fact pattern, the appropriate conclusion for the investigator to make is that the method of entry is unknown.

Another example of how an opinion is inappropriately used in an investigative report is the residential burglary. The suspect ransacks the house, turning drawers upside-down and spilling their contents on the floor, emptying closets and throwing clothes onto the furniture and floor, and leaving few things untouched. After the investigation, the investigator writes a report in which he or she reconstructs the crime and says with certainty which room the suspect entered first, which drawer was dumped out first, what items of property the suspect removed, and in what order. When questioned about how the investigator can be certain how the crime occurred, he or she will almost always say it is from experience. Experience may be a very desirable trait to have, but it is not appropriate in a factual investigative report. In this example, it would be more appropriate to say that the house was ransacked, and leave it at that.

The Expert Opinion

As mentioned earlier, in some instances, the inclusion of opinions is not only appropriate but necessary. The vast majority of these occur during arrests, when the **expert opinion** of the investigator establishes the corpus of a crime that is the basis for an arrest. An expert opinion is usually based on a combination of experience, training, and education that a particular investigator has and that qualifies him or her to offer a more credible opinion as to what has occurred. Two examples of this are arrests for being drunk and driving under the influence.

First, the situation of a common drunk requires an investigator to understand and interpret the symptoms of alcohol intoxication and evaluate these symptoms against the applicable ordinance or law prohibiting such behavior. Once the investigator has determined that a person, Steve Drinker in this instance, is violating an applicable law, he or she establishes the corpus of such a crime by forming an opinion based on all available information that Drinker is in violation and arrests him. The investigator might show this in his or her report by writing:

> Based on Drinker's objective symptoms of alcohol intoxication, I formed the opinion he was under the influence of alcohol to the point he could not care for himself or others, and I arrested him.

The second situation in which an opinion is properly used to establish the corpus of the crime is in the case of driving under the influence. In this example, during a field investigation, the investigator sees the symptoms of alcohol influence, as well as the effect of this alcohol influence on the person's ability to operate a motor vehicle in a safe and prudent manner. The investigation may include observations of improper driving or behavior that is well outside the norm for persons who are not under the influence of alcohol in similar driving conditions. Based on this driving and the obvious signs of alcohol influence, the investigator forms the opinion that the driver, Steve Drinker, is operating a motor vehicle under the influence of alcohol and arrests him for that offense. This might be written as:

> Based on Drinker's objective symptoms, his inability to operate a car, and the results of the field sobriety test, I formed the opinion that he was under the influence of an alcoholic beverage to the point that he could not operate a motor vehicle in a safe and prudent manner, and I arrested him.

DOCUMENTING RESPONSES TO MIRANDA RIGHTS

Another area in which opinions show up time and time again is when investigators write about giving the Miranda admonishment to a suspect and his or her resulting waiver. The timing and procedure to be used in giving someone her or his rights per the Miranda decision is more appropriately discussed in an investigation class; however, the matter of how to record this act is properly addressed here.

Regardless of how the matter is taken care of, it is important for the investigator to be able to accurately recall the event months later. One successful method is to read the **Miranda admonishment** to the person and quote his or her response to the two waiver questions. This could be written as:

> I read Drinker his rights per Miranda from a card I carry for that purpose. In response to the two waiver questions, he said, "Yes" and "I'll talk to you," respectively.

By quoting the responses and referring them to the waiver questions, you will remove the guesswork from the process and will be able to testify with certainty as to what was said. The problem is that many investigators forget the importance of being able to accurately recall what the suspect said and as a result fail to write it in their report. Instead of quoting the suspect, they write:

> I advised Drinker of his Miranda rights, and he acknowledged and waived them.
> I gave Drinker his rights, and he waived them in an intelligent manner.

Both of these examples have the classic ring of a good investigator to them, but they are disasters waiting to happen. In both cases, the investigator failed to establish that the Miranda admonition was given properly or that the person who received the warning understood it and actually waived his or her right against self-incrimination. What we have in this example is an investigator's opinion that the person heard the actual Miranda admonishment that he or she is entitled to and then waived those rights. What would happen if the investigator was asked to repeat in court the exact words he or she used to give this admonishment and, in so doing, forgot or left out part of it? The end result might be the exclusion of any information the person told the investigator that helped establish a crime.

It would be better to be in a position of being able to read the rights from the very same card or form that was used to admonish the suspect and then be able to quote his or her exact words. This would allow the trier of fact to decide if the admonishment was given properly and if the waiver was an intelligent and voluntary one.

It is just as important to include a negative response to the Miranda admonishment in a report because it alerts all who read the report to the fact that the suspect does not want to talk about the charges against him or her. Just as in the waiver situation, it is important to use the exact words the suspect used when invoking his or her Miranda rights. One way this might be written is:

> I read Cleworth his rights per Miranda from a card I carry for that purpose. In response to the two waiver questions he said, "Yeah" and "I'm not saying anything until I talk to my lawyer," respectively.

DOCUMENTING FIELD SHOW UPS

Still another area in which the investigator's opinion seems to creep into reports with some regularity is during **field identifications** or **show ups**. These are times when a suspect is stopped near a crime scene and a witness is brought to the suspect to see if the person being detained is or is not the perpetrator. It is important to remember that the same reasoning applies to field identifications as to Miranda admonishments. It is the investigator's job to see that the words the witness used are accurately recorded in his or her notes and later used in the report to establish probable cause or, just as important, to eliminate someone from the suspect pool. When field identifications are done, witnesses should view the suspect separately, and the admonition and response should be available for later testimony. The words the witness used should be quoted when referring to what happened, not paraphrased or generalized by the investigator. The way this might be worded in a report is:

> I stopped one block from where Brown was being detained and read Fava the field admonishment from a card I carry for that purpose. I drove to within 20 feet of Brown, and when Fava saw Brown, he said, "That's him; there is no doubt in my mind."

By reporting the facts in this manner, it allows the trier of fact to decide whether or not the witness made a good identification.

When people talk about good report writers, they are really talking about what constitutes a good report. A good report has all the right things in it and none of the wrong things. It meets the needs of those who read it yet does not require a great deal of time to complete or read.

It is important to understand who reads reports and why. A long list of persons who do can quickly be developed. It includes supervisors, bureau commanders, division commanders, chiefs of police, district attorneys, defense attorneys, defendants, planners, city administrators, newspaper reporters, and the general public. Many of the reports an investigator prepares have the chance of ending up in the hands of just about anyone, and certainly the information in the report can end up under the eyes of anyone who can read.

READER USE CONFLICT

The issue of whom a report is being written for must be recognized because the people who review it may have different needs or desires. This is the issue of **reader use conflict**, and understanding it can make the task of completing a quality report much easier.

The investigator should remember this because it can help him or her understand what needs to be done to achieve a quality investigation. An important step is to place yourself in the position of those who might be the biggest critics of your work and look at the entire situation through their eyes. Look closely at what is developing, and if a potential problem or **report writing issue** appears, take care of it by doing what is appropriate given the legal constraints, rules, and procedures you are operating under. By taking care of potential problems at that stage of the investigation, completing the report later is no problem because all you have to do is write about what happened. In most criminal investigations, there are several common issues, some of which are:

Why Are You Investigating This in the First Place?

If you recall the rule on how to begin the report, this is an easy issue to take care of. By starting the narrative with the date, time, and how you got involved, the answer is right there in the first sentence of the report.

How Did You Focus on the Suspect?

Writing in chronological order allows you to lay out, step by step, what happened and the things you did that ultimately led to the suspect. It may be the result of good investigative work, or it may be the result of good luck. Regardless, the narrative should contain nothing but the truth, and the truth should be written in factual terms.

Why Did You Detain and Search the Suspect?

Probable cause is the answer here as well. Can you write what the facts were that led you to believe it was necessary to detain the suspect and search him or her? If so, you have passed the test and eliminated this issue.

Did You Establish a Corpus?

Do you know what the corpus is for the situation? If not, can you find out by reviewing the appropriate criminal codes or jury instructions? You need to know what constitutes a crime and how to establish it in writing.

What Did You Do With the Evidence?

Did you establish a good chain of custody? Are you able to account for the evidence for every second since it was seized? Have you handled it properly? Is it marked and packaged in an acceptable manner?

Why Did You Arrest the Suspect?

This, once again, is in the area of probable cause. It is not enough to have had a good reason; you must be able to write about it so that whoever reads the report will be able to understand your reasoning.

Was Miranda Documented in the Proper Manner?

Will you be able to testify about what was said and who said it in regard to the Miranda admonishment and questions when the case reaches court? Is what you wrote free of opinions? If not, you need to rethink how to make it factual in future cases.

Self-Assessment

By placing yourself in the role of the devil's advocate, you can see where improvement is needed and work to eliminate the problems that might arise. If you are able to recognize a problem at the investigative stage and arrive at a successful way of dealing with it, the task of writing the report is nothing more than putting down on paper what you did. If you are able to eliminate the issues of report writing when they come up, you will be establishing good investigative habits. This will allow you to do a thorough job throughout your career and achieve the goal of investigating, which is to find the truth.

SUMMARY

Even when an investigator knows and follows the law, rules of evidence, and agency policy, there is still the chance that a problem might arise in the case. These potential problems are issues and can include opinions being used to establish probable cause or providing and documenting the Miranda admonishment or a witness's response during a field show up.

The investigator needs to remain true to the goal of the investigation, which is to find the truth, and not get caught up in trying to please too many taskmasters. The report should be unbiased and objective and cover what happened. A good way to avoid problems in all of these areas is to look at what you are doing through the eyes of your biggest critic and make sure you are doing the right thing. If you are, the truth will speak for itself.

REVIEW

1. Facts are things that can be proven.
2. Opinions are personal beliefs that have limited use in investigative reports.
3. Probable cause can be established by expert opinions.
4. An appropriate method of giving the Miranda admonishment is to read the admonishment and quote the person's response.
5. Issues are potential problems.
6. Different readers have different needs when it comes to what they want and expect in a report. This is known as reader use conflict.

EXERCISES

1. Read the following report, identify any issues, and discuss them in class. Rewrite the report following the rules of narrative writing. For the purposes of the report, the paramedics' names are R. Ho and G. Najjar. Captain Ho is in charge of the crew and the one you spoke to.

   Type: Death Report

   Date and Time of Occurrence: 2-22-2006 at 1500 Hours

   Date: 2-22-2006

   Location: 17221 Main Street, El Fuego

   Victim: Brown, Roger Anthony: Male, White, 49 Years

   Address: 17221 Main Street, El Fuego

   How Occurred: Liver Failure

   Discovered or Witnessed By: Brown, Marian NMN

   Address and Phone: 17221 Main St., El Fuego (123) 555-1212

   Did Ambulance Respond? No

   Doctor on Duty? No

   Victim Moved To: The Neptune Society

Moved by Whom: The Neptune Society

Relative of Victim: Marian Brown, Wife

Relative Notified? Yes

By Whom? The R/P, Marian Brown

Victim Under Care of Private Physician? Yes, 4 Years

When Private Doctor Last Seen: 1-29-2006

Name of Private Doctor: Ben Togani, MD, El Fuego Med. Center

Nature of Ailment: Cirrhosis of the Liver

Was Doctor Present? No

Victim Pronounced Dead By: El Fuego Fire Department

Death Certificate to Be Signed By: Coroner's Office

Coroner Notified By: EFPD

Describe Any Drugs Found: Lasix 80 mg., Propranolol 10 mg.

DETAILS: At approximately 1510 hours this date, I was dispatched to the Main location. Upon my arrival there, I contacted the R/P, wife of the deceased, Marian Brown.

She stated that at approximately 1440 hours, this date, she checked on her husband, who was in the upstairs master bedroom, and at that time, although he was feeling ill, he was resting in bed, dressed in a pair of white shorts. She subsequently left the location to go to the store. When she returned home at approximately 1500 hours, she went upstairs to check on her husband; at that time, she discovered him lying in a supine position, deceased. She immediately contacted El Fuego Paramedics. The El Fuego Paramedic team responded to the location after being dispatched at 1501 hours. Upon their arrival there, a fire unit pronounced the deceased dead and notified El Fuego Police Department at 1510 hours.

Sergeant Smith responded to the Main location, and after being advised by the reporting officer of the discovery and the prior physical condition of the deceased, Sergeant Smith subsequently left the location, leaving the reporting officer in charge.

I then recontacted the R/P, Mrs. Brown. She stated to me that her husband has been suffering from severe cirrhosis of the liver for approximately four years. He has been treated by numerous physicians at the Renal Clinic portion of the El Fuego Medical Center in the city of El Fuego. The last time he was at the facility, which was on January 29th, he saw Doctor Togani, who had prescribed several of the above listed medications. Mrs. Brown further indicated that on or about the 15th of January, Mr. Brown had been admitted to the El

Fuego Medical Center hospital, where he spent two days due to severe blood and potassium level depletion. She stated upon his return to his residence, he did not follow the advice of his physicians and continued drinking alcoholic beverages on a daily basis.

I then responded to the upstairs master bedroom, located upstairs in the northeast corner of the residence. There, I entered the master bedroom and observed the deceased to be lying in a supine position with his right shoulder and back against the mattress, along with the right side of his head lying against the mattress. I could observe blood emanating from the deceased nose and mouth, a small pool of which had formed on the mattress. Upon closer examination of the body, I observed severe swelling in the lower abdomen and upper stomach area of the deceased. There was no evidence of fowl play on the body and the body was still clad in a pair of white shorts.

Crime Scene Investigator James, #1207, arrived at approximately 1555 hours. He took photographs of the deceased and the surrounding bedroom area.

Upon close examination of the bedroom, hallway, and bathroom, it was apparent to the reporting officer that after the R/P, Mrs. Brown, had left the location, the deceased had used the upstairs bathroom, which is located next to the master bedroom. A small amount of blood was observed on the cabinets and on the floor of the bathroom. From there was a small spotting of the carpeting leading to the area of the bed where the deceased, after passing away, fell backwards, lying in a supine position, causing a small amount of blood to form on the bed.

Grande County Coroner's Deputy Rolly P. Wedges responded to the location at approximately 1625 hours. At that time, after close examination of the deceased and the above described evidence, he substantiated the reporting officer's conclusions. For details regarding Deputy Coroner Wedges' report, see Coroner's case number DB-95-1631.

Deputy Coroner Wedges took charge of the body and indicated that there would be no autopsy and the mortuary of the Neptune Society had been notified and would respond to take charge of the body. At that time, Deputy Wedges indicated that he would remain with the body and take charge of same until their arrival.

2. Read the following report and identify any issues and narrative writing problems. Rewrite the report following the rules of narrative writing, and address the issues previously identified. For the purposes of the exercise, assume that Linton is driving under the influence of alcohol, that it is lawful to arrest him, and that once he has been arrested, he

must provide a blood, breath, or urine sample to determine his level of intoxication.

ARRESTED:	Linton, Dwight K. DOB 5-14-63
CHARGES:	Driving under the influence
LOCATION:	Parkside and Stark, El Fuego
DATE & TIME:	3-7-07 2245 Hours
OFFICERS:	M.Colin, #4787, and R. Sproul, EA#12298
VEHICLE:	1989 Porsche 924, blue, license QUIK ONE
WITNESS 1:	Books, Linda L. DOB 6-21-65
	2555 Stark, #G, El Fuego
WITNESS 2:	Budding, Frank D. DOB 8-13-54
	2565 Stark, #K, El Fuego

DETAILS: At approximately 2239 hours, I received a radio call of a disturbance at 2555 Stark, apartment G. The call stated that the R/P's ex-husband just hit the R/P and the subject was now taking the R/P's blue 1989 Porsche 924. R/P wants to prosecute the subject who is taking her car, one Dwight K. Linton.

After receiving the above call, I arrived at the location at approximately 2245 hours. The apartment building at 2555 Stark is on the southeast corner of Stark and Parkside. As I pulled up, I parked on the northeast corner of Parkside and Stark. At that time I was approached by witness Budding and a few other people who were standing on the corner. Mr. Budding and the other people told me that they had seen a subject leave at a high rate of speed from the area and offered to be witnesses if any were needed. They advised me that something had happened directly across the street (pointing to apartment G at 2555 Stark). They told me they did not know what had gone on. I advised them that the lady at that location had called about a problem with her ex-husband or boyfriend.

I then got out of my car and was getting a clipboard for a possible stolen car report when the suspect drove the above listed blue Porsche southbound on Parkside. As he was doing so, he was also honking the horn. Mr. Budding and several other subjects then pointed the car out to me stating, "There he is." I still had not contacted the R/P prior to this time. The only thing I knew on the call was that this was apparently the ex-husband of the R/P. Seeing that the car was now evidently being returned, I walked over to contact the ex-husband (arrestee). The suspect was now parking the vehicle on Parkside, just south of Stark, on the east side. He parked his vehicle parallel at the curb just north of the driveway, which goes behind the apartment complex at 2555 Stark.

As I approached the vehicle I observed him get out of the vehicle and then close the driver door. I observed his gait to be extremely unsteady, and he was staggering and holding onto the car for balance. As I walked up to him and stood next to him, it was apparent that he did not even notice my presence (I'm standing 2 to 3 feet from him). I watched him for approximately a minute and a half attempt to put a key into the door of the vehicle and lock it. He then looked up at me, and I noticed that his eyes were bloodshot and watery and his face was flushed. He had a very strong odor of an alcoholic beverage about his breath and clothing. Not knowing the R/P's name, I then asked him if he was the ex-husband that I had been called on. He then mumbled at me words to the effect that he probably was, and I noted that his speech was extremely thick and slurred. It was very apparent to me that the suspect was drunk and was unable to care for his safety. I had also observed him just drive up in a vehicle in this condition. I then asked for his driver's license. He looked at me blankly and began to stagger toward the sidewalk. I let him walk out of the street and then took hold of his arm on the sidewalk, stopping his forward motion. I again asked him for his driver's license. He then looked at me and told me that he didn't have one. I told him that I was arresting him for driving under the influence of alcohol, at which time he mumbled something incoherently at me and began to try to walk away again. I again grabbed his arm and he began trying to twist away from me.

The suspect's balance was extremely uncoordinated, and it made him very hard to handle physically because of his size (approximately 6 feet tall, 230 pounds). I then was able to put his right arm in a wristlock behind his back and sit him down to the ground and then roll him over onto his stomach. I then placed his other hand behind his back and handcuffed him. I then escorted him (holding him up as he staggered down the sidewalk) to my police car. He was complaining to me as I walked him down the sidewalk that he was disabled and that I had hurt him. He also, as I reached the police car, asked me what I had arrested him for. I again told him that he had been arrested for driving under the influence. He asked me why, and I told him I had seen him drive up in his car. He then stated to me, "What car?" At that time I realized that his comment drew somewhat of a chuckle from a small group of people who were standing on the corner next to my police car; Mr. Budding was one of the citizens in that group. Officer Sproul, #12298, then arrived on the scene and stood by, with the suspect in my vehicle, while I contacted the R/P. The R/P (witness Books) at that time did not know that I had arrested her ex-husband and that her car had been returned.

She told me that approximately 10 minutes ago her husband had taken the car without her permission and had physically assaulted her in doing so. She told me that the police had been called to her house twice before earlier this evening regarding her ex-husband, who was drunk. She advised that both times the police responded, her husband had left the area and had not been found.

She told me she had first came home at approximately 6:30 this evening and the suspect at that time was extremely drunk. I asked her if he was drinking after she arrived home, and she replied that she had not been home most of that time but had been in and out. She did state that he had taken a gallon of rum with him and he had not returned with it prior to taking her car, and she assumed that he had been drinking it.

She advised me that her husband, just prior to taking the car, came into the house (apartment G) and had wanted her keys. She told him that she did not want him to take her car, at which time the suspect forcibly took the keys and went out to the parking area behind the apartments. She advised that the suspect got in her vehicle and she had attempted to pull the keys away from him. The suspect had backed up, with witness Books leaning in the driver's door, attempting to get the keys out. The suspect, after backing up a few feet, had pulled forward and to the left to negotiate driving out of the driveway westbound onto Parkside. The driveway is approximately 10 to 12 feet wide and has cinder block abutments on both sides. She advised that he did not negotiate his exit properly and the left door of the car struck one of the cinder block abutments and knocked it down. She stated at that time she was able to grab the keys from him and run back around the front of the apartment. She attempted to get into the apartment of a friend of hers (apartment C, downstairs from her apartment). She stated that as her friend answered the door, suspect Linton came up and grabbed the keys back from her and went back around to the car. She was unable to go after him this time but heard him leaving the location on Parkside (spinning the tires and screeching). She then called the police and I was dispatched.

I checked the victim's vehicle with her. I noted there was a small amount of scraping damage on the driver's side door. She also pointed out some more very obvious scrapes on the right side door and the right rear quarter panel of her car, which she stated the suspect had also done since he had just taken the car. It was unknown what he had scraped, and no paint transfer could be seen. She advised it was very possible that after he had retaken the keys from her and had successfully left in the car, he may have very well struck one of the other cinder block abutments because of the odd position that he had came to a stop in prior to her getting the keys. At this time, she did not want to pursue the matter of assault or auto theft. She did advise me that the suspect did have some sort of back disability from some prior condition.

During my transportation, the suspect constantly told me that he was going to sue me and my family for false arrest. He kept telling me that I had not used my red lights and therefore I could not arrest him for a "502." He told me that I could only arrest him for walking drunk because he had already parked the car and was on the sidewalk when I arrested him.

Prior to getting to the station, he also demanded a blood test. Due to his demands for a blood test, I did not initially advise him of his obligation under the implied consent law but advised him that if he did wish to take a blood test, he could. I did ask him briefly if he wished to take a breath test or a urine test instead, and he demanded that he wanted a blood test only.

Just prior to 2350 hours, I escorted the suspect into the El Fuego Jail infirmary to have a blood test taken. In the room was Jail Nurse Brown and Jailer Roland. The suspect at that time told me he wanted a cigarette or he would not submit to a blood test. I told him that there was no smoking in our jail. He then told me that he wanted a lawyer before taking any blood test, and I advised him that it was not his right. He then began pounding on the table, and it appeared as if he was going to bolt out of his chair and attack. Myself and Jailer Roland then restrained him back into his chair and told him to calm down. I told him I would read him his rights under the implied consent law if he wished to hear them. I then read verbatim from state form 13353 to the suspect. He then seemed satisfied that he had been told his rights regarding tests and then said that he now wanted to take a blood test. He then submitted to a blood test without any further incident.

At 2350 hours, Nurse Brown took two vials of blood from the suspects right inner arm. Zephiran was used as the sterilizing agent. Nurse Brown retained the blood vials for processing. Brown marked, packaged, tagged, and booked the evidence.

During the blood test, the suspect kept reiterating at me that I had not used my red lights to stop him and therefore I had no case. He then again stated that he had already parked the car and was on the sidewalk before I had arrested him and therefore I could only arrest him for drunk. He also stated words to the effect that his blood test would prove this out. After the blood test, the booking was completed without any further incidence.

End of report.

M. Colin, #4787

3. Assume you are a veteran officer and your partner, a rookie named Roger Cleworth, comes to you for advice. It seems that an arrest report he wrote was "kicked back" by the sergeant because there is a Miranda problem. You read the report and find that Cleworth did a nice job of establishing probable cause and wrote a good report with one exception, the Miranda admonishment. Cleworth wrote:

"I advised Rand of his Miranda rights, which he understood and waived."

When you talk to Cleworth, he tells you that he really read the Miranda admonishment to Rand and can tell you the exact words Rand used when he waived his rights. Cleworth looks at his notebook and shows you that Rand said, "I do" when asked if he understood his rights and "yes" when asked if he would talk. What advice would you give your partner, and how should he write this piece of his report?

QUIZ

1. Opinions should not be used in police reports.
   a. True
   b. False

2. What is a fact?

3. What is an opinion?

4. What is an issue?

5. Give an example of an issue.

6. When should an opinion be used in a police report?

7. What is reader use conflict?

8. What are two incidents in which opinions might be appropriate in a police report?

9. The chapter described eight common issues in police reports. Name four.

10. One way of minimizing issues in a report is to identify them early in the case and work to correct them during the investigation. Through whose eyes should you look at your case to identify these issues?

Appendix A

Rewrite the following report using the rules of narrative writing.

Additional Information

Your back-up officer is Jeff Guidry. You recovered the rifle and booked it for safekeeping.

Details:

On 6-2-06, at approx. 2245 hrs. I discovered a subject, (Bayes, Harold Lee, 3-3-69, Driver Lic. B0527346, 590 Gilmar, Duncanville, 735-2848) asleep in a vehicle parked in the parking lot W/of Palo Verde and S/of Spring. The vehicle was a Toyota, 2002, Tan 4X4 pick-up w/camper shell, lic# 33R162L. I also could see the wooden stock and blue receiver of a rifle lying next to him.

After getting follow-up officer at the scene I was able to get the subject Bayes out of the vehicle. He said that he had recently been kicked out of his residence and had come to the park to job but fell asleep afterwards inside his vehicle. He said that the rifle was just part of his property which he'd taken with him and had no other reason for placing the rifle next to him. The weapon, (Remington, M-740, 30.6, SER # 3822162) was siezed from the vehicle and found to have (1) expended cartridge in the chamber. No other ammunition was found.

An attempt was made to have Duncanville PD check his residence however they were too busy at the time. They did go by the residence later and were unable to contact anyone. (They also advised that they'd been out to Subject Bayes residence last night on a call of an attempted suicide where Bayes was the victim/suspect. It is unknown if my contact was related).

Bayes was F.I.'d and released at the scene. His rifle was siezed and booked at El Fuego PD Property Room for safekeeping.

Rewrite the following report using the rules of narrative writing.

Additional Information

You received a radio call to meet the reporting party, Katherine Tunney, at 1230 hours on 7-25-06. Katherine Tunney lives at home with her parents, Carol and Gary Tunney. The term 28 is jargon for radio code 10-28, which is a request to check the registration of a license plate.

Information Report

Location: 1647 Rancho, El Fuego
Occurrence Date: 7-25-06
Other Person: Tunney, Carol P. DOB 1-21-61
Other Person: Tunney, Katherine L. DOB 9-16-84

Details:

I made contact with Katherine Tunney who stated that at approximately 1215 hours this date, she was getting ready to leave her residence to do errands in the area when she observed a white Toyota van wagon to be stopped in front of her residence on the opposite side of the street facing westbound. What brought her attention to the vehicle was the fact that the driver, who appeared to be the sole occupant of the vehicle was turned in the front driver's seat with a camera up to his face taking pictures of her specific residence. Miss Tunney indicated that the vehicle, although parked facing westbound, was stopped in the middle of the street and not on the complete opposite side.

Miss Tunney indicated that she thought this was rather suspicious, remembering her mother had been contacted by a suspicious subject back in April of this year.

She responded out to make contact with the subject who, upon seeing her exit the residence, immediately rolled up his driver's window and sped away at a fast rate of speed to apparently vacate the area. Miss Tunney indicated she was able to obtain the license number of the van when it slowed for a speed bump which was located close to the residence. I asked Miss Tunney for that license number and she indicated it was 836L92F.

A 28 was run on this license plate and it came back to a 1984 Toyota Wagon, registered to Bond, Stephen D., 3381 El Balazo, El Fuego. Miss Tunney describes the male subject to be a male white, with graying hair, NFD.

In speaking with Mrs. Tunney (Carol) she had indicated she filed a report with the police department back in April of this year. She was out in the front of her garage, in the driveway repotting a plant when she was approached by a suspicious subject who asked what her name was. Mrs. Tunney (Carol) had indicated that she told the subject she wasn't going to tell him her name and asked what his name was. Apparently she did not get any information from that.

The subject left in a vehicle which Mrs. Tunney stated had a personalized plate of ISPYONU. A 28 information was run on that vehicle. It comes back to a 2005 SAAB registered to Toomey, Richard, 36 Paseo De Bonito, El Rancho.

I advised Miss Tunney and Mrs. Tunney that at this time we did not have a crime that had occurred. However, this information report would be filed. At this time, no contact has been made with the driver of the Toyota or the Saab and information has been passed on to the investigation division.

EXERCISE 3

Rewrite the following report using the rules of narrative writing.

ARREST REPORT

Arrested:	Jones, Scott A. DOB 1-1-85
Charges:	Grand Theft Auto
Location:	Beach and Yorktown, El Fuego
Date and Time:	6-3-06 0340 Hours
Officers:	Pete Martin and Jim Rust

Details:

On 6-3-06 at approximately 0320 hours, Officer Martin and myself were dispatched to the Stop and Rob Market at Beach and Yorktown regarding a subject who had called the police department and stated he had stolen a motorcycle and wanted to give himself up. According to dispatch, the subject sounded somewhat intoxicated on the telephone.

As Officer Martin and myself turned onto Yorktown from Beach, we observed a subject standing by a public telephone urinating on the sidewalk. As we pulled into the parking lot, the subject turned and faced us and began zipping up his fly with his hand. At that time we exited our vehicle and made contact with the subject who identified himself as Jones, Scott. When I asked him what the problem was he stated he had stolen a motorcycle from a friend and wanted to turn himself in. It should be noted that while I was talking with Jones he appeared to be under the influence of alcohol, but not totally intoxicated. His eyes were somewhat bloodshot and watery and his speech was a little bit slurred, but not too bad. I could also detect an odor of alcohol on his breath. He was also slightly injured, as there were abrasions on his hands, elbows, forehead and chin. When I asked him how he got these abrasions he stated he had crashed on the motorcycle that he had stolen. I asked him where the motorcycle was and he stated he had parked it down the street in the bushes. I then asked him who the motorcycle belonged to and he stated it belonged to a friend of his by the name of Dave. When I asked him where Dave resides, he stated Dave was not at home and that he was in jail.

At this time based on our investigation, we advised Jones of his constitutional rights per Miranda in order to further conduct an investigation and he stated yes to every question except #1 and #2. Jones was then taken into custody, handcuffed and placed him in the rear of our unit and we then proceeded to go westbound on Yorktown. Upon approaching the intersection of Yorktown and Garfield, Jones stated he had thrown the motorcycle into the bushes indicating the bushes on the southeast corner of the intersection. Upon checking this location we did observe the motorcycle lying on its side in the bushes. There was no registration on the motorcycle, however there was vehicle identification that we had run through our Communications Center. While out in the field we were advised the stolen vehicle system was down as well as the registration system and it was unknown ETA when it would be available.

In further conversation, I asked again who the motorcycle belonged to and Scott stated Dave, however Dave had been arrested apparently "Last Thursday" for riding on unimproved property by the El Fuego Police Department. Scott was unable to provide us with Dave's last name and said he did not have a phone number nor could he provide us with an address to where we could verify whether or not the bike was stolen. In further conversation he changed his story that he did not steal the bike from Dave but had stolen it from his girlfriend, Pam. Again he was unable to provide us with the last name nor could he provide us with a phone number or address where Pam could be contacted.

Based on our conversation with Scott and him being somewhat intoxicated and urinating in public, he was transported to our jail facility and he was booked for suspicion of auto theft. While we were booking Scott I again contacted our Communications Center and asked them to try running the VIN number again to determine the owner. They stated upon running the VIN number, it came back the vehicle had been reported stolen and they also indicated the registration information on file showed the owner is Dave Rogers, 616 Elm, El Fuego and the Case number on the stolen was 06-13821. The vehicle was impounded and transported to BEST Towing where it was stored.

Appendix B

EL SEGUNDO POLICE DEPARTMENT

SUSPECT REPORT

PAGE_____ OF _____

CASE NO

CRIME 1				
CODE SECTION	CRIME	CLASSIFICATION		REFER OTHER REPORTS
LOCATION (Be Specific)		RD.	DATE	TIME	SUPPL. ☐	INCIDENT NO.

SUSP. VEH. 2

LICENSE #	STATE	VEH. YR	MAKE	MODEL	BODY STYLE

BODY STYLE: ☐ 0 UNK ☐ 2 4-DR ☐ 4 P/U ☐ 6 VAN ☐ 8 RV ☐ 10 OTHER
☐ 1 2-DR ☐ 3 CONV ☐ 5 TRUCK ☐ 7 S/W ☐ 9 M/C

COLOR/COLOR OTHER CHARACTERISTICS (i.e. T/C Damage, Unique Marks or Paint, etc.) DISPOSITION OF VEH.

REGISTERED OWNER

SUSPECT 3

SUSP. # NAME (First, Last, Middle)

SEX
☐ 1. M
☐ 2. F

RACE
☐ 0 UNK ☐ 2 HISP ☐ 4 IND ☐ 6 JAP ☐ 8 OTH_____
☐ 1 WHT ☐ 3 BLK ☐ 5 CHI ☐ 7 FIL ☐ 9 P.ISL.

AKA

D.O.B. AGE HT. WT. BUILD
☐ 1 THIN ☐ 3 HEAVY
☐ 0 UNK ☐ 2 MED ☐ 4 MUSCLR

HAIR
☐ 0 UNK ☐ 2 BLK ☐ 4 RED ☐ 6 S/P ☐ 8 OTHER
☐ 1 BRN ☐ 3 BLN ☐ 5 GRAY ☐ 7 WHT

EYES
☐ 0 UNK ☐ 2 BLK ☐ 4 GRN ☐ 6 GRAY
☐ 1 BRN ☐ 3 BLU ☐ 5 HAZEL ☐ 7 OTHER

D.L. #

RES. ADDRESS RD ZIP CODE RES. PHONE # (   ) S.S #

BUS. ADDRESS RD ZIP CODE BUS. PHONE # (   ) OCCUPATION

CLOTHING

ARRESTED
☐ 1 YES ☐ 2 NO

STATUS
☐ 1 DRIVER ☐ 3 PED.
☐ 2 PASS

GANG AFFILIATION:
HOW KNOWN:
☐ 1 KNOWN
☐ 2 SUSPECTED

AMT. OF HAIR 4	HAIR STYLE 8	COMPLEXION 10	TATTOOS/SCARS 13	DISTING. MARKS 14	WEAPON(S) 17
☐ 0 UNKNOWN Q21	☐ 0 UNKNOWN Q25	☐ 0 UNKNOWN Q27	☐ 0 UNKNOWN	☐ 0 NONE Q30	☐ 0 UNKNOWN ☐ 0 NONE Q33
☐ 1 THICK	☐ 1 LONG	☐ 1 CLEAR	☐ 1 FACE	_____	☐ 1 CLUB _____
☐ 2 THIN	☐ 2 SHORT	☐ 2 ACNE	☐ 2 TEETH	_____	☐ 2 HAND GUN _____
☐ 3 RECEDING	☐ 3 COLLAR	☐ 3 POCKED	☐ 3 NECK	_____	☐ 3 OTHER UNK GUN _____
☐ 4 BALD	☐ 4 MILITARY	☐ 4 FRECKLED	☐ 4 R/ARM	_____	☐ 4 RIFLE _____
☐ 5 OTHER_____	☐ 5 CREW CUT	☐ 5 WEATHERED	☐ 5 L/ARM	_____	☐ 5 SHOT GUN _____
TYPE OF HAIR 5	☐ 6 RIGHT PART	☐ 6 ALBINO	☐ 6 R/HAND	_____	☐ 6 TOY GUN _____
☐ 0 UNKNOWN Q22	☐ 7 LEFT PART	☐ 7 OTHER_____	☐ 7 L/HAND	_____	☐ 7 SIMULATED _____
☐ 1 STRAIGHT	☐ 8 CENTER PART	GLASSES 11	☐ 8 R/LEG	_____	☐ 8 POCKET KNIFE _____
☐ 2 CURLY	☐ 9 STRAIGHT BACK	☐ 0 UNKNOWN Q28	☐ 9 L/LEG	_____	☐ 9 BUTCHER KNIFE _____
☐ 3 WAVY	☐ 10 PONY TAIL	☐ 0 NONE	☐ 10 R/SHOULDER	_____	☐ 10 OTH. CUT/STAB INST _____
☐ 4 FINE	☐ 11 AFRO/NATURAL	☐ 1 YES (No Descrip.)	☐ 11 L/SHOULDER	_____	☐ 11 HANDS/FEET _____
☐ 5 COARSE	☐ 12 PROCESSED	☐ 2 REG GLASSES	☐ 12 FRONT TORSO	_____	☐ 12 BODILY FORCE _____
☐ 6 WIRY	☐ 13 TEASED	☐ 3 SUN GLASSES	☐ 13 BACK TORSO	_____	☐ 13 STRANGULATION _____
☐ 7 WIG	☐ 14 OTHER_____	☐ 4 WIRE FRAME	☐ 14 OTHER	_____	☐ 14 TIRE IRON _____
☐ 8 OTHER_____	FACIAL HAIR 9	☐ 5 PLASTIC FRAME		_____	☐ 15 OTHER _____
HAIR CONDITION 6	☐ 0 UNKNOWN Q26	☐ Color_____			WEAPON FEATURE 18
☐ 0 UNKNOWN Q23	☐ 0 N/A	☐ 6 OTHER_____	UNIQUE CLTHNG 15	WEAPON IN 16	☐ 0 UNKNOWN ☐ 0 NONE Q34
☐ 1 CLEAN	☐ 1 CLN SHAVEN	VOICE 12	☐ 0 UNK ☐ 0 NONE	☐ 0 UNKNOWN Q32	☐ 1 ALTERED STOCK _____
☐ 2 DIRTY	☐ 2 MOUSTACHE	☐ 0 UNKNOWN Q29	☐ 1 CAP/HAT Q31	☐ 0 N/A	☐ 2 SAWED OFF _____
☐ 3 GREASY	☐ 3 FULL BEARD	☐ 0 N/A	_____	☐ 1 BAG/BRIEFCASE	☐ 3 AUTOMATIC _____
☐ 4 MATTED	☐ 4 GOATEE	☐ 1 LISP	☐ 2 GLOVES	☐ 2 NEWSPAPER	☐ 4 BOLT ACTION _____
☐ 5 ODOR	☐ 5 FUMANCHU	☐ 2 SLURRED	_____	☐ 3 POCKET	☐ 5 PUMP _____
☐ 6 OTHER_____	☐ 6 LOWER LIP	☐ 3 STUTTER	☐ 3 SKI MASK	☐ 4 SHOULDER	☐ 6 REVOLVER _____
R/L HANDED 7	☐ 7 SIDE BURNS	☐ 4 ACCENT		HOLSTER	☐ 7 BLUE STEEL _____
☐ 0 UNKNOWN Q24	☐ 8 FUZZ	Describe_____	☐ 4 STOCKING MASK	☐ 5 WAISTBAND	☐ 8 CHROME/NICKEL _____
☐ 1 RIGHT	☐ 9 UNSHAVEN	_____		☐ 6 OTHER_____	☐ 9 DOUBLE BARREL _____
☐ 2 LEFT	☐ 10 OTHER_____	☐ 5 OTHER_____	☐ 5 OTHER_____		☐ 10 SINGLE BARREL _____
					☐ 11 OTHER _____

REPORTING OFFICER ID# DATE REVIEWED BY ID# DATE

COPIES: ☐ CHIEF ☐ CII ☐ PATROL ☐ DB ☐ OTHER AGENCY ROUTED BY ENTERED BY
TO: ☐ ☐ CAU ☐ ABC (2 copies) ☐ DA _____

ESPD Form #200 (Rev 5/97)

El Segundo Police Department Suspect Report

		CASE NO.
		PAGE _____

SUSPECT · **3**

SUSP. #	NAME (First, Last, Middle)	SEX □ 1 M □ 2 F	RACE	□ 0 UNK □ 2 HISP □ 4 IND □ 6 JAP □ 8 OTH_____ □ 1 WHT □ 3 BLK □ 5 CHI □ 7 FIL □ 9 P.ISL.

AKA	D.O.B.	AGE	HT.	WT.	BUILD

BUILD: □ 1 THIN   □ 3 HEAVY / □ 0 UNK   □ 2 MED   □ 4 MUSCLR

HAIR	□ 0 UNK □ 2 BLK □ 4 RED □ 6 S/P □ 8 OTHER □ 1 BRN □ 3 BLN □ 5 GRAY □ 7 WHT _____	EYES	□ 0 UNK □ 2 BLK □ 4 GRN □ 6 GRAY □ 1 BRN □ 3 BLU □ 5 HAZEL □ 7 OTHER ____	D.L. #

RES. ADDRESS	RD	ZIP CODE	RES. PHONE # ( )	S.S #

BUS. ADDRESS (School)	RD	ZIP CODE	BUS. PHONE # ( )	OCCUPATION

CLOTHING	ARRESTED □ 1 YES □ 2 NO	STATUS □ 1 DRIVER □ 3 PED. □ 2 PASS	GANG AFFILIATION: HOW KNOWN:	□ 1 KNOWN □ 2 SUSPECTED

AMT. OF HAIR **4**	HAIR STYLE **8**	COMPLEXION **10**	TATTOOS/SCARS **13**	DISTING. MARKS **14**	WEAPON(S) **17**
□ 0 UNKNOWN	□ 0 UNKNOWN	□ 0 UNKNOWN	□ 0 UNKNOWN	□ 0 NONE	□ 0 UNKNOWN □ 0 NONE
□ 1 THICK	□ 1 LONG	□ 1 CLEAR	□ 1 FACE	_____	□ 1 CLUB _____
□ 2 THIN	□ 2 SHORT	□ 2 ACNE	□ 2 TEETH	_____	□ 2 HAND GUN _____
□ 3 RECEDING	□ 3 COLLAR	□ 3 POCKED	□ 3 NECK	_____	□ 3 OTHER UNK GUN _____
□ 4 BALD	□ 4 MILITARY	□ 4 FRECKLED	□ 4 R/ARM	_____	□ 4 RIFLE _____
□ 5 OTHER_____	□ 5 CREW CUT	□ 5 WEATHERED	□ 5 L/ARM	_____	□ 5 SHOT GUN _____
TYPE OF HAIR **5**	□ 6 RIGHT PART	□ 6 ALBINO	□ 6 R/HAND	_____	□ 6 TOY GUN _____
□ 0 UNKNOWN	□ 7 LEFT PART	□ 7 OTHER_____	□ 7 L/HAND	_____	□ 7 SIMULATED _____
□ 1 STRAIGHT	□ 8 CENTER PART	GLASSES **11**	□ 8 R/LEG	_____	□ 8 POCKET KNIFE _____
□ 2 CURLY	□ 9 STRAIGHT BACK	□ 0 UNKNOWN	□ 9 L/LEG	_____	□ 9 BUTCHER KNIFE _____
□ 3 WAVY	□ 10 PONY TAIL	□ 0 NONE	□ 10 R/SHOULDER	_____	□ 10 OTH. CUT/STAB INST _____
□ 4 FINE	□ 11 AFRO/NATURAL	□ 1 YES (No Descrip.)	□ 11 L/SHOULDER	_____	□ 11 HANDS/FEET _____
□ 5 COARSE	□ 12 PROCESSED	□ 2 REG GLASSES	□ 12 FRONT TORSO	_____	□ 12 BODILY FORCE _____
□ 6 WIRY	□ 13 TEASED	□ 3 SUN GLASSES	□ 13 BACK TORSO	_____	□ 13 STRANGULATION _____
□ 7 WIG	□ 14 OTHER_____	□ 4 WIRE FRAME	□ 14 OTHER	_____	□ 14 TIRE IRON _____
□ 8 OTHER_____	FACIAL HAIR **9**	□ 5 PLASTIC FRAME		_____	□ 15 OTHER _____
HAIR CONDITION **6**	□ 0 UNKNOWN	□ Color_____			WEAPON FEATURE **18**
□ 0 UNKNOWN	□ 0 N/A	□ 6 OTHER_____	UNIQUE CLTHNG **15**	WEAPON IN **16**	□ 0 UNKNOWN □ 0 NONE
□ 1 CLEAN	□ 1 CLN SHAVEN	VOICE **12**	□ 0 UNK □ 0 NONE	□ 0 UNKNOWN	□ 1 ALTERED STOCK _____
□ 2 DIRTY	□ 2 MOUSTACHE	□ 0 UNKNOWN	□ 1 CAP/HAT	□ 0 N/A	□ 2 SAWED OFF _____
□ 3 GREASY	□ 3 FULL BEARD	□ 0 N/A	_____	□ 1 BAG/BRIEFCASE	□ 3 AUTOMATIC _____
□ 4 MATTED	□ 4 GOATEE	□ 1 LISP	□ 2 GLOVES	□ 2 NEWSPAPER	□ 4 BOLT ACTION _____
□ 5 ODOR	□ 5 FUMANCHU	□ 2 SLURRED		□ 3 POCKET	□ 5 PUMP _____
□ 6 OTHER_____	□ 6 LOWER LIP	□ 3 STUTTER	□ 3 SKI MASK	□ 4 SHOULDER	□ 6 REVOLVER _____
R/L HANDED **7**	□ 7 SIDE BURNS	□ 4 ACCENT	_____	HOLSTER	□ 7 BLUE STEEL _____
□ 0 UNKNOWN	□ 8 FUZZ	Describe_____	□ 4 STOCKING MASK	□ 5 WAISTBAND	□ 8 CHROME/NICKEL _____
□ 1 RIGHT	□ 9 UNSHAVEN	_____		□ 6 OTHER_____	□ 9 DOUBLE BARREL _____
□ 2 LEFT	□ 10 OTHER_____	□ 5 OTHER_____	□ 5 OTHER_____	_____	□ 10 SINGLE BARREL _____
					□ 11 OTHER _____

_____

_____

_____

_____

_____

_____

_____

_____

_____

ESPD Form #200 (Rev 5/97)

(Continued)

EL SEGUNDO POLICE DEPARTMENT

SUSPECT REPORT

PAGE_____ OF _____

CASE NO

CRIME 1

CODE SECTION	CRIME	CLASSIFICATION				REFER OTHER REPORTS

LOCATION (Be Specific)		RD.	DATE	TIME	SUPPL. ☐	INCIDENT NO.

SUSP. VEH 2

LICENSE #	STATE	VEH. YR	MAKE	MODEL	BODY STYLE

BODY STYLE: ☐ 0 UNK ☐ 2 4-DR ☐ 4 P/U ☐ 6 VAN ☐ 8 RV ☐ 10 OTHER
☐ 1 2-DR ☐ 3 CONV ☐ 5 TRUCK ☐ 7 S/W ☐ 9 M/C

COLOR/COLOR	OTHER CHARACTERISTICS (i.e. T/C Damage, Unique Marks or Paint, etc.)	DISPOSITION OF VEH.

REGISTERED OWNER

SUSPECT 3

SUSP. #	NAME (First, Last, Middle)	SEX	RACE

SEX: ☐ 1. M ☐ 2. F

RACE: ☐ 0 UNK ☐ 2 HISP ☐ 4 IND ☐ 6 JAP ☐ 8 OTH_____
☐ 1 WHT ☐ 3 BLK ☐ 5 CHI ☐ 7 FIL ☐ 9 P.ISL.

AKA	D.O.B.	AGE	HT.	WT.	BUILD

BUILD: ☐ 0 UNK ☐ 1 THIN ☐ 2 MED ☐ 3 HEAVY ☐ 4 MUSCLR

HAIR: ☐ 0 UNK ☐ 2 BLK ☐ 4 RED ☐ 6 S/P ☐ 8 OTHER _____
☐ 1 BRN ☐ 3 BLN ☐ 5 GRAY ☐ 7 WHT _____

EYES: ☐ 0 UNK ☐ 2 BLK ☐ 4 GRN ☐ 6 GRAY
☐ 1 BRN ☐ 3 BLU ☐ 5 HAZEL ☐ 7 OTHER _____

D.L. #

RES. ADDRESS	RD	ZIP CODE	RES. PHONE # ( )	S.S #

BUS. ADDRESS	RD	ZIP CODE	BUS. PHONE # ( )	OCCUPATION

CLOTHING	ARRESTED	STATUS	GANG AFFILIATION:

ARRESTED: ☐ 1 YES ☐ 2 NO

STATUS: ☐ 1 DRIVER ☐ 3 PED. ☐ 2 PASS

HOW KNOWN: ☐ 1 KNOWN ☐ 2 SUSPECTED

AMT. OF HAIR 4	HAIR STYLE 8	COMPLEXION 10	TATTOOS/SCARS 13	DISTING. MARKS 14	WEAPON(S) 17
☐ 0 UNKNOWN Q21	☐ 0 UNKNOWN Q25	☐ 0 UNKNOWN Q27	☐ 0 UNKNOWN	☐ 0 NONE Q30	☐ 0 UNKNOWN ☐ 0 NONE Q33
☐ 1 THICK	☐ 1 LONG	☐ 1 CLEAR	☐ 1 FACE _____		☐ 1 CLUB _____
☐ 2 THIN	☐ 2 SHORT	☐ 2 ACNE	☐ 2 TEETH _____		☐ 2 HAND GUN _____
☐ 3 RECEDING	☐ 3 COLLAR	☐ 3 POCKED	☐ 3 NECK _____		☐ 3 OTHER UNK GUN _____
☐ 4 BALD	☐ 4 MILITARY	☐ 4 FRECKLED	☐ 4 R/ARM _____		☐ 4 RIFLE _____
☐ 5 OTHER_____	☐ 5 CREW CUT	☐ 5 WEATHERED	☐ 5 L/ARM _____		☐ 5 SHOT GUN _____
TYPE OF HAIR 5	☐ 6 RIGHT PART	☐ 6 ALBINO	☐ 6 R/HAND _____		☐ 6 TOY GUN _____
☐ 0 UNKNOWN Q22	☐ 7 LEFT PART	☐ 7 OTHER_____	☐ 7 L/HAND _____		☐ 7 SIMULATED _____
☐ 1 STRAIGHT	☐ 8 CENTER PART	**GLASSES 11**	☐ 8 R/LEG _____		☐ 8 POCKET KNIFE _____
☐ 2 CURLY	☐ 9 STRAIGHT BACK	☐ 0 UNKNOWN Q28	☐ 9 L/LEG _____		☐ 9 BUTCHER KNIFE _____
☐ 3 WAVY	☐ 10 PONY TAIL	☐ 0 NONE	☐ 10 R/SHOULDER _____		☐ 10 OTH. CUT/STAB INST _____
☐ 4 FINE	☐ 11 AFRO/NATURAL	☐ 1 YES (No Descrip.)	☐ 11 L/SHOULDER _____		☐ 11 HANDS/FEET _____
☐ 5 COARSE	☐ 12 PROCESSED	☐ 2 REG GLASSES	☐ 12 FRONT TORSO _____		☐ 12 BODILY FORCE _____
☐ 6 WIRY	☐ 13 TEASED	☐ 3 SUN GLASSES	☐ 13 BACK TORSO _____		☐ 13 STRANGULATION _____
☐ 7 WIG	☐ 14 OTHER_____	☐ 4 WIRE FRAME	☐ 14 OTHER _____		☐ 14 TIRE IRON _____
☐ 8 OTHER_____	**FACIAL HAIR 9**	☐ 5 PLASTIC FRAME	_____		☐ 15 OTHER _____
HAIR CONDITION 6	☐ 0 UNKNOWN Q26	☐ Color_____	_____		**WEAPON FEATURE 18**
☐ 0 UNKNOWN Q23	☐ 0 N/A	☐ 6 OTHER_____	**UNIQUE CLTHNG 15**	**WEAPON IN 16**	☐ 0 UNKNOWN ☐ 0 NONE Q34
☐ 1 CLEAN	☐ 1 CLN SHAVEN	**VOICE 12**	☐ 0 UNK ☐ 0 NONE	☐ 0 UNKNOWN Q32	☐ 1 ALTERED STOCK _____
☐ 2 DIRTY	☐ 2 MOUSTACHE	☐ 0 UNKNOWN Q29	☐ 1 CAP/HAT Q31	☐ 0 N/A	☐ 2 SAWED OFF _____
☐ 3 GREASY	☐ 3 FULL BEARD	☐ 0 N/A	_____	☐ 1 BAG/BRIEFCASE	☐ 3 AUTOMATIC _____
☐ 4 MATTED	☐ 4 GOATEE	☐ 1 LISP	☐ 2 GLOVES	☐ 2 NEWSPAPER	☐ 4 BOLT ACTION _____
☐ 5 ODOR	☐ 5 FUMANCHU	☐ 2 SLURRED		☐ 3 POCKET	☐ 5 PUMP _____
☐ 6 OTHER_____	☐ 6 LOWER LIP	☐ 3 STUTTER	☐ 3 SKI MASK	☐ 4 SHOULDER HOLSTER	☐ 6 REVOLVER _____
R/L HANDED 7	☐ 7 SIDE BURNS	☐ 4 ACCENT	_____		☐ 7 BLUE STEEL _____
☐ 0 UNKNOWN Q24	☐ 8 FUZZ	Describe_____	☐ 4 STOCKING MASK	☐ 5 WAISTBAND	☐ 8 CHROME/NICKEL _____
☐ 1 RIGHT	☐ 9 UNSHAVEN	_____	_____	☐ 6 OTHER_____	☐ 9 DOUBLE BARREL _____
☐ 2 LEFT	☐ 10 OTHER_____	☐ 5 OTHER_____	☐ 5 OTHER _____		☐ 10 SINGLE BARREL _____
					☐ 11 OTHER _____

REPORTING OFFICER	ID#	DATE	REVIEWED BY	ID#	DATE

COPIES: ☐ CHIEF ☐ CII ☐ PATROL ☐ DB ☐ OTHER AGENCY
TO: ☐ ☐ CAU ☐ ABC (2 copies) ☐ DA _____

ROUTED BY | ENTERED BY

ESPD Form #200 (Rev 5/97)

El Segundo Police Department Suspect Report

		CASE NO.	
		PAGE _____	

SUSPECT — 3

SUSP. #	NAME (First, Last, Middle)	SEX ☐ 1 M ☐ 2 F	RACE ☐ 0 UNK ☐ 2 HISP ☐ 4 IND ☐ 6 JAP ☐ 8 OTH_____ ☐ 1 WHT ☐ 3 BLK ☐ 5 CHI ☐ 7 FIL ☐ 9 P.ISL.

AKA		D.O.B.	AGE	HT.	WT.	BUILD ☐ 1 THIN ☐ 3 HEAVY ☐ 0 UNK ☐ 2 MED ☐ 4 MUSCLR

HAIR ☐ 0 UNK ☐ 2 BLK ☐ 4 RED ☐ 6 S/P ☐ 8 OTHER ☐ 1 BRN ☐ 3 BLN ☐ 5 GRAY ☐ 7 WHT _____	EYES ☐ 0 UNK ☐ 2 BLK ☐ 4 GRN ☐ 6 GRAY ☐ 1 BRN ☐ 3 BLU ☐ 5 HAZEL ☐ 7 OTHER ____	D.L. #

RES. ADDRESS	RD	ZIP CODE	RES. PHONE # ( )	S.S #

BUS. ADDRESS (School)	RD	ZIP CODE	BUS. PHONE # ( )	OCCUPATION

CLOTHING	ARRESTED ☐ 1 YES ☐ 2 NO	STATUS ☐ 1 DRIVER ☐ 3 PED. ☐ 2 PASS	GANG AFFILIATION: HOW KNOWN:	☐ 1 KNOWN ☐ 2 SUSPECTED

AMT. OF HAIR 4	HAIR STYLE 8	COMPLEXION 10	TATTOOS/SCARS 13	DISTING. MARKS 14	WEAPON(S) 17
☐ 0 UNKNOWN	☐ 0 UNKNOWN	☐ 0 UNKNOWN	☐ 0 UNKNOWN ☐ 0 NONE		☐ 0 UNKNOWN ☐ 0 NONE
☐ 1 THICK	☐ 1 LONG	☐ 1 CLEAR	☐ 1 FACE	_____	☐ 1 CLUB _____
☐ 2 THIN	☐ 2 SHORT	☐ 2 ACNE	☐ 2 TEETH	_____	☐ 2 HAND GUN _____
☐ 3 RECEDING	☐ 3 COLLAR	☐ 3 POCKED	☐ 3 NECK	_____	☐ 3 OTHER UNK GUN _____
☐ 4 BALD	☐ 4 MILITARY	☐ 4 FRECKLED	☐ 4 R/ARM	_____	☐ 4 RIFLE _____
☐ 5 OTHER_____	☐ 5 CREW CUT	☐ 5 WEATHERED	☐ 5 L/ARM	_____	☐ 5 SHOT GUN _____
TYPE OF HAIR 5	☐ 6 RIGHT PART	☐ 6 ALBINO	☐ 6 R/HAND	_____	☐ 6 TOY GUN _____
☐ 0 UNKNOWN	☐ 7 LEFT PART	☐ 7 OTHER_____	☐ 7 L/HAND	_____	☐ 7 SIMULATED _____
☐ 1 STRAIGHT	☐ 8 CENTER PART	**GLASSES 11**	☐ 8 R/LEG	_____	☐ 8 POCKET KNIFE _____
☐ 2 CURLY	☐ 9 STRAIGHT BACK	☐ 0 UNKNOWN	☐ 9 L/LEG	_____	☐ 9 BUTCHER KNIFE _____
☐ 3 WAVY	☐ 10 PONY TAIL	☐ 0 NONE	☐ 10 R/SHOULDER	_____	☐ 10 OTH. CUT/STAB INST _____
☐ 4 FINE	☐ 11 AFRO/NATURAL	☐ 1 YES (No Descrip.)	☐ 11 L/SHOULDER	_____	☐ 11 HANDS/FEET _____
☐ 5 COARSE	☐ 12 PROCESSED	☐ 2 REG GLASSES	☐ 12 FRONT TORSO	_____	☐ 12 BODILY FORCE _____
☐ 6 WIRY	☐ 13 TEASED	☐ 3 SUN GLASSES	☐ 13 BACK TORSO	_____	☐ 13 STRANGULATION _____
☐ 7 WIG	☐ 14 OTHER_____	☐ 4 WIRE FRAME	☐ 14 OTHER	_____	☐ 14 TIRE IRON _____
☐ 8 OTHER_____	**FACIAL HAIR 9**	☐ 5 PLASTIC FRAME	_____		☐ 15 OTHER _____
HAIR CONDITION 6	☐ 0 UNKNOWN	☐ Color_____			**WEAPON FEATURE 18**
☐ 0 UNKNOWN	☐ 0 N/A	☐ 6 OTHER_____	**UNIQUE CLTHNG 15**	**WEAPON IN 16**	☐ 0 UNKNOWN ☐ 0 NONE
☐ 1 CLEAN	☐ 1 CLN SHAVEN	**VOICE 12**	☐ 0 UNK ☐ 0 NONE	☐ 0 UNKNOWN	☐ 1 ALTERED STOCK _____
☐ 2 DIRTY	☐ 2 MOUSTACHE	☐ 0 UNKNOWN	☐ 1 CAP/HAT	☐ 0 N/A	☐ 2 SAWED OFF _____
☐ 3 GREASY	☐ 3 FULL BEARD	☐ 0 N/A	_____	☐ 1 BAG/BRIEFCASE	☐ 3 AUTOMATIC _____
☐ 4 MATTED	☐ 4 GOATEE	☐ 1 LISP	☐ 2 GLOVES	☐ 2 NEWSPAPER	☐ 4 BOLT ACTION _____
☐ 5 ODOR	☐ 5 FUMANCHU	☐ 2 SLURRED	_____	☐ 3 POCKET	☐ 5 PUMP _____
☐ 6 OTHER_____	☐ 6 LOWER LIP	☐ 3 STUTTER	☐ 3 SKI MASK	☐ 4 SHOULDER	☐ 6 REVOLVER _____
R/L HANDED 7	☐ 7 SIDE BURNS	☐ 4 ACCENT	_____	HOLSTER	☐ 7 BLUE STEEL _____
☐ 0 UNKNOWN	☐ 8 FUZZ	Describe_____	☐ 4 STOCKING MASK	☐ 5 WAISTBAND	☐ 8 CHROME/NICKEL _____
☐ 1 RIGHT	☐ 9 UNSHAVEN	_____	_____	☐ 6 OTHER_____	☐ 9 DOUBLE BARREL _____
☐ 2 LEFT	☐ 10 OTHER_____	☐ 5 OTHER_____	☐ 5 OTHER_____	_____	☐ 10 SINGLE BARREL _____
					☐ 11 OTHER _____

_____

_____

_____

_____

_____

_____

_____

_____

ESPD Form #200 (Rev 5/97)

(Continued)

EL SEGUNDO POLICE DEPARTMENT
PROPERTY REPORT

DATE AND TIME OF REPORT

DR#

- ☐ ARRESTEE
- ☐ SUSPECT
- ☐ VICTIM
- ☐ WITNESS

- ☐ EVIDENCE
- ☐ FOUND PROPERTY
- ☐ SAFEKEEPING
- ☐ PRISONER PROPERTY

- ☐ DESTRUCTION
- ☐ UNDER OBSERVATION
- ☐ HOLD: CONTACT OFFICER _____
- ☐ OTHER _____

NAME (Property Booked To)

NAME (Secondary Persons)

☐ MISDEMEANOR
☐ FELONY

ADDRESS

ADDRESS

CHARGES

CITY   ZIP

CITY   ZIP

1.

PHONE

PHONE

2.

RES.:   BUS.:

RES.:   BUS.

3.

B I K E	MAKE	MODEL	SPEED ☐ BOY'S 1 ☐ 3 ☐ ☐ GIRL'S 5☐ 10 ☐ 15 ☐ OTHER ☐	FRAME COLOR	WHEEL SIZE
	SERIAL NO.	LICENSE NO. CITY	OTHER DESCRIPTIONS		

FOUND PROPERTY STATEMENT: DO YOU WISH TO CLAIM THIS PROPERTY IF THE LEGAL OWNER CANNOT BE LOCATED?   ☐ YES   ☐ NO
PROPERTY CAN BE CLAIMED AFTER **90 DAYS** DATE........................   FINDER SIGNATURE..

PRISONER PROPERTY STATEMENT: ITEMS NOT ABLE TO GO TO COURT WILL BE HELD BY THIS DEPARTMENT FOR A PERIOD OF **ONE YEAR.** IF AFTER 1 YEAR YOU HAVE NOT MADE ARRANGEMENTS WITH THE PROPERTY & EVIDENCE OFFICER TO CLAIM YOUR PROPERTY WE WILL DESTROY ALL PRISONER PROPERTY AFTER THAT DATE.
PRISONER SIGNATURE .... DATE...............................

LIST ALL ARTICLES STARTING WITH ITEM NO. 1: DESCRIBE PROPERTY AS FOLLOWS.

ITEM NO.	QTY.	ARTICLE, DESCRIPTION, BRAND NAME, MODEL, ETC.	SERIAL NO.	VALUE OR WEIGHT

CONTINUATION FORM USED FOR ADDITIONAL ITEMS ☐

COMPLAINT FORM NARRATION:

REPORTING OFFICER (Name and Serial No.)	TEMPORARY LOCATION OF PROPERTY	APPROVING SUPERVISOR	DATE
PROPERTY USE ONLY PROPERTY ROOM LOCATION	BY:	DATE	FINAL DISPOSITION
PROPERTY RELEASED TO:	ADDRESS:		DATE

WHITE: WITH EVIDENCE   **YELLOW:** TO RECORDS   **PINK:** TO FINDER/PRISONER PROPERTY

El Segundo Police Department Property Report

CONTINUATION REPORT

DATE			OFFICER		DR#		
ITEM NO.	QTY.	ARTICLE DESCRIPTION, BRAND NAME, MODEL ETC.			SERIAL NO.	VALUE	

REPORTING OFFICER (Name and Serial No.)	TEMPORARY LOCATION OF PROPERTY	APPROVING SUPERVISOR	DATE
PROPERTY USE ONLY PROPERTY ROOM LOCATION	BY:	DATE	FINAL DISPOSITION
FCN NO.	BY:		DATE

(Continued))

EL SEGUNDO POLICE DEPARTMENT
PROPERTY REPORT

DATE AND TIME OF REPORT

DR#

☐ ARRESTEE
☐ SUSPECT
☐ VICTIM
☐ WITNESS

☐ EVIDENCE
☐ FOUND PROPERTY
☐ SAFEKEEPING
☐ PRISONER PROPERTY

☐ DESTRUCTION
☐ UNDER OBSERVATION
☐ HOLD: CONTACT OFFICER _____
☐ OTHER _____

NAME (Property Booked To)

NAME (Secondary Persons)

☐ MISDEMEANOR
☐ FELONY

ADDRESS

ADDRESS

CHARGES
1.

CITY   ZIP

CITY   ZIP

PHONE

PHONE

2.

RES.:   BUS.:

RES.:   BUS.

3.

| B I K E | MAKE | MODEL | ☐ BOY'S ☐ GIRL'S SPEED 1 ☐ 3 ☐ 5☐ 10 ☐ 15 ☐ OTHER ☐ | FRAME COLOR | WHEEL SIZE |
| | SERIAL NO. | LICENSE NO.   CITY | OTHER DESCRIPTIONS | | |

FOUND PROPERTY STATEMENT: DO YOU WISH TO CLAIM THIS PROPERTY IF THE LEGAL OWNER CANNOT BE LOCATED?   ☐ YES   ☐ NO
PROPERTY CAN BE CLAIMED AFTER **90 DAYS** DATE........................ FINDER SIGNATURE..

PRISONER PROPERTY STATEMENT: ITEMS NOT ABLE TO GO TO COURT WILL BE HELD BY THIS DEPARTMENT FOR A PERIOD OF **ONE YEAR.** IF AFTER 1 YEAR YOU HAVE NOT MADE
ARRANGEMENTS WITH THE PROPERTY & EVIDENCE OFFICER TO CLAIM YOUR PROPERTY WE WILL DESTROY ALL PRISONER PROPERTY AFTER THAT DATE.
PRISONER SIGNATURE .... DATE.............................

LIST ALL ARTICLES STARTING WITH ITEM NO. 1: DESCRIBE PROPERTY AS FOLLOWS.

ITEM NO.	QTY.	ARTICLE, DESCRIPTION, BRAND NAME, MODEL, ETC.	SERIAL NO.	VALUE OR WEIGHT

CONTINUATION FORM USED FOR ADDITIONAL ITEMS ☐

COMPLAINT FORM NARRATION:

REPORTING OFFICER (Name and Serial No.)	TEMPORARY LOCATION OF PROPERTY	APPROVING SUPERVISOR	DATE
PROPERTY USE ONLY PROPERTY ROOM LOCATION	BY:	DATE	FINAL DISPOSITION
PROPERTY RELEASED TO:	ADDRESS:		DATE

WHITE: WITH EVIDENCE   **YELLOW:** TO RECORDS   **PINK:** TO FINDER/PRISONER PROPERTY

El Segundo Police Department Property Report

CONTINUATION REPORT

DATE			OFFICER			DR#		

ITEM NO.	QTY.	ARTICLE DESCRIPTION, BRAND NAME, MODEL ETC.	SERIAL NO.	VALUE

REPORTING OFFICER (Name and Serial No.)	TEMPORARY LOCATION OF PROPERTY	APPROVING SUPERVISOR	DATE
PROPERTY USE ONLY PROPERTY ROOM LOCATION	BY:	DATE	FINAL DISPOSITION
FCN NO.	BY:		DATE

(Continued)

EL SEGUNDO POLICE DEPARTMENT
PROPERTY REPORT

DATE AND TIME OF REPORT

DR#

☐ ARRESTEE
☐ SUSPECT
☐ VICTIM
☐ WITNESS

☐ EVIDENCE
☐ FOUND PROPERTY
☐ SAFEKEEPING
☐ PRISONER PROPERTY

☐ DESTRUCTION
☐ UNDER OBSERVATION
☐ HOLD: CONTACT OFFICER _____
☐ OTHER _____

NAME (Property Booked To)

NAME (Secondary Persons)

☐ MISDEMEANOR
☐ FELONY

ADDRESS

ADDRESS

CHARGES

1.

CITY   ZIP

CITY   ZIP

2.

PHONE

PHONE

RES.:   BUS.:

RES.:   BUS.

3.

B I K E	MAKE	MODEL	☐ BOY'S ☐ GIRL'S SPEED 1☐ 3☐ 5☐ 10☐ 15☐ OTHER☐	FRAME COLOR	WHEEL SIZE
	SERIAL NO.	LICENSE NO. CITY	OTHER DESCRIPTIONS		

FOUND PROPERTY STATEMENT: DO YOU WISH TO CLAIM THIS PROPERTY IF THE LEGAL OWNER CANNOT BE LOCATED?   ☐ YES   ☐ NO
PROPERTY CAN BE CLAIMED AFTER **90 DAYS** DATE......................... FINDER SIGNATURE...

PRISONER PROPERTY STATEMENT: ITEMS NOT ABLE TO GO TO COURT WILL BE HELD BY THIS DEPARTMENT FOR A PERIOD OF **ONE YEAR.** IF AFTER 1 YEAR YOU HAVE NOT MADE ARRANGEMENTS WITH THE PROPERTY & EVIDENCE OFFICER TO CLAIM YOUR PROPERTY WE WILL DESTROY ALL PRISONER PROPERTY AFTER THAT DATE.
PRISONER SIGNATURE .... DATE.........................

LIST ALL ARTICLES STARTING WITH ITEM NO. 1: DESCRIBE PROPERTY AS FOLLOWS.

ITEM NO.	QTY.	ARTICLE, DESCRIPTION, BRAND NAME, MODEL, ETC.	SERIAL NO.	VALUE OR WEIGHT

CONTINUATION FORM USED FOR ADDITIONAL ITEMS ☐

COMPLAINT FORM NARRATION:

REPORTING OFFICER (Name and Serial No.)	TEMPORARY LOCATION OF PROPERTY	APPROVING SUPERVISOR	DATE
PROPERTY USE ONLY PROPERTY ROOM LOCATION	BY:	DATE	FINAL DISPOSITION
PROPERTY RELEASED TO:	ADDRESS:		DATE

WHITE: WITH EVIDENCE   **YELLOW:** TO RECORDS   **PINK:** TO FINDER/PRISONER PROPERTY

El Segundo Police Department Property Report

CONTINUATION REPORT

DATE		OFFICER			DR#	

ITEM NO.	QTY.	ARTICLE DESCRIPTION, BRAND NAME, MODEL ETC.	SERIAL NO.	VALUE

REPORTING OFFICER (Name and Serial No.)	TEMPORARY LOCATION OF PROPERTY	APPROVING SUPERVISOR	DATE
PROPERTY USE ONLY PROPERTY ROOM LOCATION	BY:	DATE	FINAL DISPOSITION
FCN NO.	BY:		DATE

(Continued)

La Habra Police Department Evidence Tag (Front)

La Habra Police Department Evidence of Tag (Reverse)

La Habra Police Department Evidence Tag (Front)

La Habra Police Department Evidence Tag (Reverse)

☐ NO PROSECUTION DESIRED	**EL SEGUNDO POLICE DEPARTMENT**	A ☐ ACTIVE	CASE NUMBER
☐ TELEPHONE REPORT	348 MAIN STREET	S ☐ SUSPENDED	
☐ INSURANCE REPORT	EL SEGUNDO, CA 90245	R ☐ RECORDS	
☐ COURTESY REPORT	310-322-9114	C ☐ CLOSED	REFER OTHER RPTS
☐ DOMESTIC VIOLENCE	**CRIME REPORT**	K ☐ COURTESY	
☐ CONFIDENTIAL SEX CRIME		U ☐ UNFOUNDED	

CRIME

CODE SECTION	CRIME			UCR CODE	SECONDARY-COUNTS	OTHER-COUNTS
SPECIFIC LOCATION OF CRIME		OCCURRED ON/OR BETWEEN:	DATE	DAY	TIME	
BUSINESS NAME	DATE RPT'D	TIME RPT'D	AND:	DATE	DAY	TIME

VICTIM

NAME (Last, First, Middle)	OCCUPATION	D.O.B.	AGE	SEX ☐ 1. M ☐ 2. F	RACE ☐ 1. WHT ☐ 2. HISP ☐ 3. BLK ☐ 5. CHI ☐ 7. FIL ☐ 9. P.ISL ☐ 4. IND ☐ 6. JAP ☐ 8. OTH.
RESIDENCE ADDRESS	CITY	ZIP CODE	RES. PHONE ( )		
BUSINESS NAME AND ADDRESS	CITY	ZIP CODE	BUS. PHONE ( )		

VICTIM(S) - WITNESS - RP

CODE	NAME (Last, First, Middle)	OCCUPATION	D.O.B.	AGE	SEX ☐ 1. M ☐ 2. F	RACE ☐ 1. WHT ☐ 2. HISP ☐ 3. BLK ☐ 5. CHI ☐ 7. FIL ☐ 9. P.ISL ☐ 4. IND ☐ 6. JAP ☐ 8. OTH.
	RESIDENCE ADDRESS	CITY	ZIP CODE	RES. PHONE ( )		
	BUSINESS NAME AND ADDRESS	CITY	ZIP CODE	BUS. PHONE ( )		
CODE	NAME (Last, First, Middle)	OCCUPATION	D.O.B.	AGE	SEX ☐ 1. M ☐ 2. F	RACE ☐ 1. WHT ☐ 2. HISP ☐ 3. BLK ☐ 5. CHI ☐ 7. FIL ☐ 9. P.ISL ☐ 4. IND ☐ 6. JAP ☐ 8. OTH.
	RESIDENCE ADDRESS	CITY	ZIP CODE	RES. PHONE ( )		
	BUSINESS NAME AND ADDRESS	CITY	ZIP CODE	BUS. PHONE ( )		
CODE	NAME (Last, First, Middle)	OCCUPATION	D.O.B.	AGE	SEX ☐ 1. M ☐ 2. F	RACE ☐ 1. WHT ☐ 2. HISP ☐ 3. BLK ☐ 5. CHI ☐ 7. FIL ☐ 9. P.ISL ☐ 4. IND ☐ 6. JAP ☐ 8. OTH.
	RESIDENCE ADDRESS	CITY	ZIP CODE	RES. PHONE ( )		
	BUSINESS ADDRESS	CITY	ZIP CODE	BUS. PHONE ( )		

VIC VEH

LICENSE #	STATE	YEAR	MAKE	MODEL	BODY STYLE ☐ 0 UNK ☐ 2 4-DR ☐ 4 P/U ☐ 6 VAN ☐ 8 RV ☐ 10 OTHER ☐ 1 2-DR ☐ 3 CONV ☐ 5 TRUCK ☐ 7 S/W ☐ 9 M/C
COLOR/COLOR	OTHER CHARACTERISTICS (i.e. T/C Damage, Unique Marks or Paint, etc.)			DISPOSITION OF VEHICLE	

FACTORS

- ☐ 1 THERE IS A WITNESS TO THE CRIME   SUSPECT PAGE ☐ YES ☐ NO
- ☐ 2 A SUSPECT WAS ARRESTED
- ☐ 3 A SUSPECT WAS NAMED
- ☐ 4 A SUSPECT CAN BE LOCATED
- ☐ 5 A SUSPECT CAN BE DESCRIBED
- ☐ 6 A SUSPECT CAN BE IDENTIFIED
- ☐ 7 A SUSPECT VEHICLE CAN BE IDENTIFIED
- ☐ 8 THERE IS IDENTIFIABLE STOLEN PROPERTY
- ☐ 9 THERE IS A SIGNIFIANT M.O.
- ☐ 10 SIGNIFICANT PHYSICAL EVIDENCE IS PRESENT
- ☐ 11 THERE IS A MAJOR INJURY/SEX CRIME INVOLVED
- ☐ 12 THERE IS A GOOD POSSIBILITY OF A SOLUTION
- ☐ 13 FURTHER INVESTIGATION NEEDED
- ☐ 14 CRIME IS GANG RELATED
- ☐ 15 HATE CRIME RELATED

EVIDENCE

- ☐ 0 NONE
- ☐ 1 FINGERPRINTS
- ☐ 2 TOOLS
- ☐ 3 TOOL MARKINGS
- ☐ 4 GLASS
- ☐ 5 PAINT
- ☐ 6 BULLET CASING
- ☐ 7 BULLET
- ☐ 8 RAPE KIT
- ☐ 9 SEMEN
- ☐ 10 BLOOD
- ☐ 11 URINE
- ☐ 12 HAIR
- ☐ 13 FIREARMS
- ☐ 14 PHOTOGRAPHS
- ☐ 15 OTHER (DESCRIBE)

VICTIMS SIGNATURE	DATE	DETECTIVE ASSIGNED SIGNATURE	DATE

REPORTING OFFICER	ID#	DATE	REVIEWING SUPERVISOR	ID#	DATE

COPIES: ☐ CHIEF ☐ CII ☐ PATROL ☐ DB ☐ OTHER	ROUTED BY	ENTERED BY
TO: ☐ DMV ☐ CAU ☐ ABC (2 copies) ☐ DA		

ESPD Form 3/97

El Segundo Police Department Crime Report (Front and Reverse)

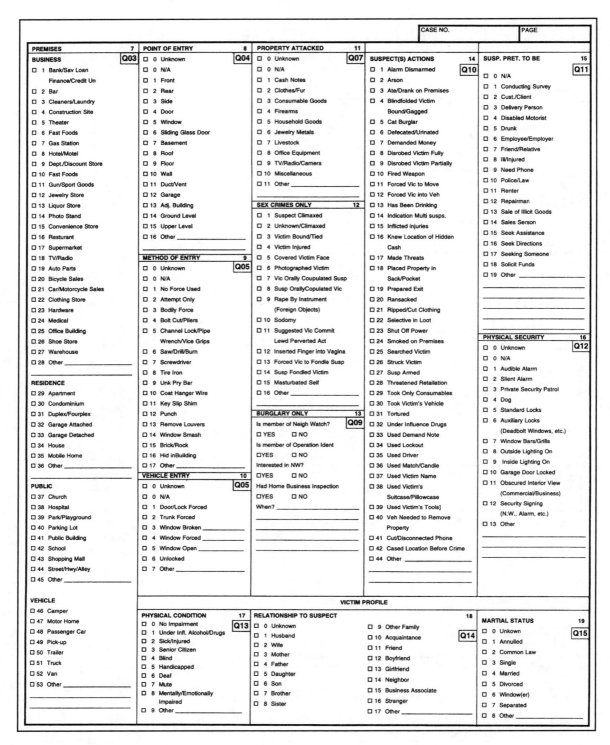

		CASE NO.		PAGE	

PREMISES 7

BUSINESS Q03
- ☐ 1 Bank/Sav Loan Finance/Credit Un
- ☐ 2 Bar
- ☐ 3 Cleaners/Laundry
- ☐ 4 Construction Site
- ☐ 5 Theater
- ☐ 6 Fast Foods
- ☐ 7 Gas Station
- ☐ 8 Hotel/Motel
- ☐ 9 Dept./Discount Store
- ☐ 10 Fast Foods
- ☐ 11 Gun/Sport Goods
- ☐ 12 Jewelry Store
- ☐ 13 Liquor Store
- ☐ 14 Photo Stand
- ☐ 15 Convenience Store
- ☐ 16 Resturant
- ☐ 17 Supermarket
- ☐ 18 TV/Radio
- ☐ 19 Auto Parts
- ☐ 20 Bicycle Sales
- ☐ 21 Car/Motorcycle Sales
- ☐ 22 Clothing Store
- ☐ 23 Hardware
- ☐ 24 Medical
- ☐ 25 Office Building
- ☐ 26 Shoe Store
- ☐ 27 Warehouse
- ☐ 28 Other _____

RESIDENCE
- ☐ 29 Apartment
- ☐ 30 Condominium
- ☐ 31 Duplex/Fourplex
- ☐ 32 Garage Attached
- ☐ 33 Garage Detached
- ☐ 34 House
- ☐ 35 Mobile Home
- ☐ 36 Other _____

PUBLIC
- ☐ 37 Church
- ☐ 38 Hospital
- ☐ 39 Park/Playground
- ☐ 40 Parking Lot
- ☐ 41 Public Building
- ☐ 42 School
- ☐ 43 Shopping Mall
- ☐ 44 Street/Hwy/Alley
- ☐ 45 Other _____

VEHICLE
- ☐ 46 Camper
- ☐ 47 Motor Home
- ☐ 48 Passenger Car
- ☐ 49 Pick-up
- ☐ 50 Trailer
- ☐ 51 Truck
- ☐ 52 Van
- ☐ 53 Other _____

POINT OF ENTRY 8 Q04
- ☐ 0 N/A
- ☐ 1 Front
- ☐ 2 Rear
- ☐ 3 Side
- ☐ 4 Door
- ☐ 5 Window
- ☐ 6 Sliding Glass Door
- ☐ 7 Basement
- ☐ 8 Roof
- ☐ 9 Floor
- ☐ 10 Wall
- ☐ 11 Duct/Vent
- ☐ 12 Garage
- ☐ 13 Adj. Building
- ☐ 14 Ground Level
- ☐ 15 Upper Level
- ☐ 16 Other _____

METHOD OF ENTRY 9 Q05
- ☐ 0 Unknown
- ☐ 0 N/A
- ☐ 1 No Force Used
- ☐ 2 Attempt Only
- ☐ 3 Bodily Force
- ☐ 4 Bolt Cut/Pilers
- ☐ 5 Channel Lock/Pipe Wrench/Vice Grips
- ☐ 6 Saw/Drill/Burn
- ☐ 7 Screwdriver
- ☐ 8 Tire Iron
- ☐ 9 Unk Pry Bar
- ☐ 10 Coat Hanger Wire
- ☐ 11 Key Slip Shim
- ☐ 12 Punch
- ☐ 13 Remove Louvers
- ☐ 14 Window Smash
- ☐ 15 Brick/Rock
- ☐ 16 Hid inBuilding
- ☐ 17 Other _____

VEHICLE ENTRY 10 Q05
- ☐ 0 Unknown
- ☐ 0 N/A
- ☐ 1 Door/Lock Forced
- ☐ 2 Trunk Forced
- ☐ 3 Window Broken _____
- ☐ 4 Window Forced _____
- ☐ 5 Window Open _____
- ☐ 6 Unlocked
- ☐ 7 Other _____

PROPERTY ATTACKED 11 Q07
- ☐ 0 Unknown
- ☐ 0 N/A
- ☐ 1 Cash Notes
- ☐ 2 Clothes/Fur
- ☐ 3 Consumable Goods
- ☐ 4 Firearms
- ☐ 5 Household Goods
- ☐ 6 Jewelry Metals
- ☐ 7 Livestock
- ☐ 8 Office Equipment
- ☐ 9 TV/Radio/Camera
- ☐ 10 Miscellaneous
- ☐ 11 Other _____

SEX CRIMES ONLY 12
- ☐ 1 Suspect Climaxed
- ☐ 2 Unknown/Climaxed
- ☐ 3 Victim Bound/Tied
- ☐ 4 Victim Injured
- ☐ 5 Covered Victim Face
- ☐ 6 Photographed Victim
- ☐ 7 Vic Orally Coupulated Susp
- ☐ 8 Susp OrallyCoupulated Vic
- ☐ 9 Rape By Instrument (Foreign Objects)
- ☐ 10 Sodomy
- ☐ 11 Suggested Vic Commit Lewd Perverted Act
- ☐ 12 Inserted Finger into Vagina
- ☐ 13 Forced Vic to Fondle Susp
- ☐ 14 Susp Fondled Victim
- ☐ 15 Masturbated Self
- ☐ 16 Other _____

BURGLARY ONLY 13 Q09
- Is member of Neigh Watch?
- ☐ YES   ☐ NO
- Is member of Operation Ident
- ☐ YES   ☐ NO
- Interested in NW?
- ☐ YES   ☐ NO
- Had Home Business Inspection
- ☐ YES   ☐ NO
- When? _____
- _____
- _____
- _____

SUSPECT(S) ACTIONS 14
- ☐ 1 Alarm Dismarmed Q10
- ☐ 2 Arson
- ☐ 3 Ate/Drank on Premises
- ☐ 4 Blindfolded Victim Bound/Gagged
- ☐ 5 Cat Burglar
- ☐ 6 Defecated/Urinated
- ☐ 7 Demanded Money
- ☐ 8 Disrobed Victim Fully
- ☐ 9 Disrobed Victim Partially
- ☐ 10 Fired Weapon
- ☐ 11 Forced Vic to Move
- ☐ 12 Forced Vic into Veh
- ☐ 13 Has Been Drinking
- ☐ 14 Indication Multi susps.
- ☐ 15 Inflicted injuries
- ☐ 16 Knew Location of Hidden Cash
- ☐ 17 Made Threats
- ☐ 18 Placed Property in Sack/Pocket
- ☐ 19 Prepared Exit
- ☐ 20 Ransacked
- ☐ 21 Ripped/Cut Clothing
- ☐ 22 Selective in Loot
- ☐ 23 Shut Off Power
- ☐ 24 Smoked on Premises
- ☐ 25 Searched Victim
- ☐ 26 Struck Victim
- ☐ 27 Susp Armed
- ☐ 28 Threatened Retaliation
- ☐ 29 Took Only Consumables
- ☐ 30 Took Victim's Vehicle
- ☐ 31 Tortured
- ☐ 32 Under Influence Drugs
- ☐ 33 Used Demand Note
- ☐ 34 Used Lockout
- ☐ 35 Used Driver
- ☐ 36 Used Match/Candle
- ☐ 37 Used Victim Name
- ☐ 38 Used Victim's Suitcase/Pillowcase
- ☐ 39 Used Victim's Tools]
- ☐ 40 Veh Needed to Remove Property
- ☐ 41 Cut/Disconnected Phone
- ☐ 42 Cased Location Before Crime
- ☐ 44 Other _____

SUSP. PRET. TO BE 15 Q11
- ☐ 0 N/A
- ☐ 1 Conducting Survey
- ☐ 2 Cust./Client
- ☐ 3 Delivery Person
- ☐ 4 Disabled Motorist
- ☐ 5 Drunk
- ☐ 6 Employee/Employer
- ☐ 7 Friend/Relative
- ☐ 8 Ill/Injured
- ☐ 9 Need Phone
- ☐ 10 Police/Law
- ☐ 11 Renter
- ☐ 12 Repairman
- ☐ 13 Sale of Illicit Goods
- ☐ 14 Sales Serson
- ☐ 15 Seek Assistance
- ☐ 16 Seek Directions
- ☐ 17 Seeking Someone
- ☐ 18 Solicit Funds
- ☐ 19 Other _____
- _____
- _____
- _____
- _____

PHYSICAL SECURITY 16 Q12
- ☐ 0 Unknown
- ☐ 0 N/A
- ☐ 1 Audible Alarm
- ☐ 2 Slient Alarm
- ☐ 3 Private Security Patrol
- ☐ 4 Dog
- ☐ 5 Standard Locks
- ☐ 6 Auxiliary Locks (Deadbolt Windows, etc.)
- ☐ 7 Window Bars/Grills
- ☐ 8 Outside Lighting On
- ☐ 9 Inside Lighting On
- ☐ 10 Garage Door Locked
- ☐ 11 Obscured Interior View (Commercial/Business)
- ☐ 12 Security Signing (N.W., Alarm, etc.)
- ☐ 13 Other _____

VICTIM PROFILE

PHYSICAL CONDITION 17 Q13
- ☐ 0 No Impairment
- ☐ 1 Under Infl. Alcohol/Drugs
- ☐ 2 Sick/Injured
- ☐ 3 Senior Citizen
- ☐ 4 Blind
- ☐ 5 Handicapped
- ☐ 6 Deaf
- ☐ 7 Mute
- ☐ 8 Mentally/Emotionally Impaired
- ☐ 9 Other _____

RELATIONSHIP TO SUSPECT 18 Q14
- ☐ 0 Unknown
- ☐ 1 Husband
- ☐ 2 Wife
- ☐ 3 Mother
- ☐ 4 Father
- ☐ 5 Daughter
- ☐ 6 Son
- ☐ 7 Brother
- ☐ 8 Sister
- ☐ 9 Other Family
- ☐ 10 Acquaintance
- ☐ 11 Friend
- ☐ 12 Boyfriend
- ☐ 13 Girlfriend
- ☐ 14 Neighbor
- ☐ 15 Business Associate
- ☐ 16 Stranger
- ☐ 17 Other _____

MARTIAL STATUS 19 Q15
- ☐ 0 Unknown
- ☐ 1 Annulled
- ☐ 2 Common Law
- ☐ 3 Single
- ☐ 4 Married
- ☐ 5 Divorced
- ☐ 6 Window(er)
- ☐ 7 Separated
- ☐ 8 Other _____

(Continued)

☐ NO PROSECUTION DESIRED	**EL SEGUNDO POLICE DEPARTMENT**		A ☐ ACTIVE	CASE NUMBER
☐ TELEPHONE REPORT	348 MAIN STREET		S ☐ SUSPENDED	
☐ INSURANCE REPORT	EL SEGUNDO, CA 90245		R ☐ RECORDS	
☐ COURTESY REPORT	310-322-9114		C ☐ CLOSED	REFER OTHER RPTS
☐ DOMESTIC VIOLENCE	**CRIME REPORT**		K ☐ COURTESY	
☐ CONFIDENTIAL SEX CRIME			U ☐ UNFOUNDED	

CRIME

CODE SECTION	CRIME			UCR CODE	SECONDARY-COUNTS	OTHER-COUNTS
SPECIFIC LOCATION OF CRIME			OCCURRED ON/OR BETWEEN:	DATE	DAY	TIME
BUSINESS NAME		DATE RPT'D TIME RPT'D	AND:	DATE	DAY	TIME

VICTIM

NAME (Last, First, Middle)	OCCUPATION	D.O.B.	AGE	SEX ☐ 1. M ☐ 2. F	RACE ☐ 1. WHT ☐ 2. HISP ☐ 3. BLK ☐ 5. CHI ☐ 7. FIL ☐ 9. P.ISL ☐ 4. IND ☐ 6. JAP ☐ 8. OTH.____
RESIDENCE ADDRESS		CITY	ZIP CODE	RES. PHONE ( )	
BUSINESS NAME AND ADDRESS		CITY	ZIP CODE	BUS. PHONE ( )	

VICTIM(S) - WITNESS - RP

CODE	NAME (Last, First, Middle)	OCCUPATION	D.O.B.	AGE	SEX ☐ 1. M ☐ 2. F	RACE ☐ 1. WHT ☐ 2. HISP ☐ 3. BLK ☐ 5. CHI ☐ 7. FIL ☐ 9. P.ISL ☐ 4. IND ☐ 6. JAP ☐ 8. OTH.____
RESIDENCE ADDRESS			CITY	ZIP CODE	RES. PHONE ( )	
BUSINESS NAME AND ADDRESS			CITY	ZIP CODE	BUS. PHONE ( )	
CODE	NAME (Last, First, Middle)	OCCUPATION	D.O.B.	AGE	SEX ☐ 1. M ☐ 2. F	RACE ☐ 1. WHT ☐ 2. HISP ☐ 3. BLK ☐ 5. CHI ☐ 7. FIL ☐ 9. P.ISL ☐ 4. IND ☐ 6. JAP ☐ 8. OTH.____
RESIDENCE ADDRESS			CITY	ZIP CODE	RES. PHONE ( )	
BUSINESS NAME AND ADDRESS			CITY	ZIP CODE	BUS. PHONE ( )	
CODE	NAME (Last, First, Middle)	OCCUPATION	D.O.B.	AGE	SEX ☐ 1. M ☐ 2. F	RACE ☐ 1. WHT ☐ 2. HISP ☐ 3. BLK ☐ 5. CHI ☐ 7. FIL ☐ 9. P.ISL ☐ 4. IND ☐ 6. JAP ☐ 8. OTH.____
RESIDENCE ADDRESS			CITY	ZIP CODE	RES. PHONE ( )	
BUSINESS ADDRESS			CITY	ZIP CODE	BUS. PHONE ( )	

VIC VEH

LICENSE #	STATE	YEAR	MAKE	MODEL	BODY STYLE ☐ 0 UNK ☐ 2 4-DR ☐ 4 P/U ☐ 6 VAN ☐ 8 RV ☐ 10 OTHER ☐ 1 2-DR ☐ 3 CONV ☐ 5 TRUCK ☐ 7 S/W ☐ 9 M/C
COLOR/COLOR		OTHER CHARACTERISTICS (i.e. T/C Damage, Unique Marks or Paint, etc.)			DISPOSITION OF VEHICLE

FACTORS

- ☐ 1 THERE IS A WITNESS TO THE CRIME   SUSPECT PAGE ☐ YES ☐ NO
- ☐ 2 A SUSPECT WAS ARRESTED
- ☐ 3 A SUSPECT WAS NAMED
- ☐ 4 A SUSPECT CAN BE LOCATED
- ☐ 5 A SUSPECT CAN BE DESCRIBED
- ☐ 6 A SUSPECT CAN BE IDENTIFIED
- ☐ 7 A SUSPECT VEHICLE CAN BE IDENTIFIED
- ☐ 8 THERE IS IDENTIFIABLE STOLEN PROPERTY
- ☐ 9 THERE IS A SIGNIFIANT M.O.
- ☐ 10 SIGNIFICANT PHYSICAL EVIDENCE IS PRESENT
- ☐ 11 THERE IS A MAJOR INJURY/SEX CRIME INVOLVED
- ☐ 12 THERE IS A GOOD POSSIBILITY OF A SOLUTION
- ☐ 13 FURTHER INVESTIGATION NEEDED
- ☐ 14 CRIME IS GANG RELATED
- ☐ 15 HATE CRIME RELATED

EVIDENCE

- ☐ 0 NONE
- ☐ 1 FINGERPRINTS
- ☐ 2 TOOLS
- ☐ 3 TOOL MARKINGS
- ☐ 4 GLASS
- ☐ 5 PAINT
- ☐ 6 BULLET CASING
- ☐ 7 BULLET
- ☐ 8 RAPE KIT
- ☐ 9 SEMEN
- ☐ 10 BLOOD
- ☐ 11 URINE
- ☐ 12 HAIR
- ☐ 13 FIREARMS
- ☐ 14 PHOTOGRAPHS
- ☐ 15 OTHER (DESCRIBE)

VICTIMS SIGNATURE	DATE	DETECTIVE ASSIGNED SIGNATURE	DATE

REPORTING OFFICER	ID#	DATE	REVIEWING SUPERVISOR	ID#	DATE

COPIES:	☐ CHIEF ☐ CII	☐ PATROL	☐ DB ☐ OTHER	ROUTED BY	ENTERED BY
TO:	☐ DMV ☐ CAU	☐ ABC (2 copies)	☐ DA		

ESPD Form 3/97

El Segundo Police Department Crime Report (Front and Reverse)

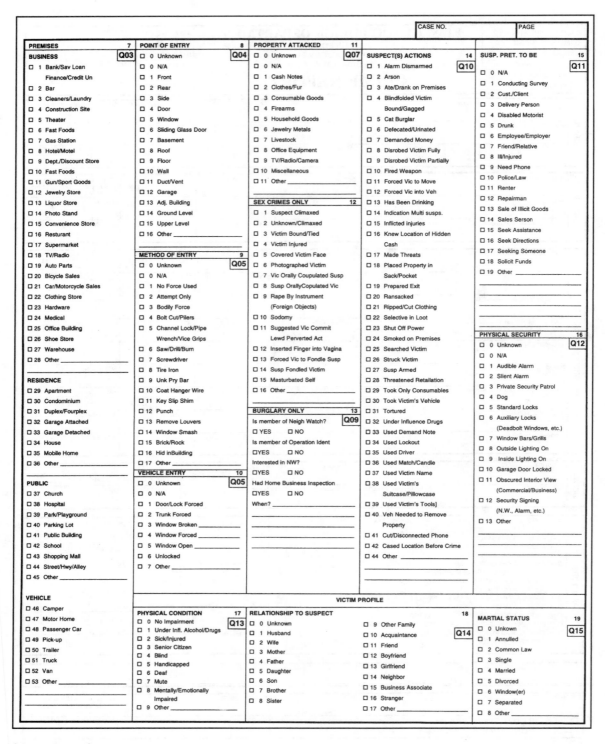

CASE NO.		PAGE	

PREMISES 7 — **Q03**

BUSINESS

- ☐ 1 Bank/Sav Loan Finance/Credit Un
- ☐ 2 Bar
- ☐ 3 Cleaners/Laundry
- ☐ 4 Construction Site
- ☐ 5 Theater
- ☐ 6 Fast Foods
- ☐ 7 Gas Station
- ☐ 8 Hotel/Motel
- ☐ 9 Dept./Discount Store
- ☐ 10 Fast Foods
- ☐ 11 Gun/Sport Goods
- ☐ 12 Jewelry Store
- ☐ 13 Liquor Store
- ☐ 14 Photo Stand
- ☐ 15 Convenience Store
- ☐ 16 Resturant
- ☐ 17 Supermarket
- ☐ 18 TV/Radio
- ☐ 19 Auto Parts
- ☐ 20 Bicycle Sales
- ☐ 21 Car/Motorcycle Sales
- ☐ 22 Clothing Store
- ☐ 23 Hardware
- ☐ 24 Medical
- ☐ 25 Office Building
- ☐ 26 Shoe Store
- ☐ 27 Warehouse
- ☐ 28 Other _____

RESIDENCE

- ☐ 29 Apartment
- ☐ 30 Condominium
- ☐ 31 Duplex/Fourplex
- ☐ 32 Garage Attached
- ☐ 33 Garage Detached
- ☐ 34 House
- ☐ 35 Mobile Home
- ☐ 36 Other _____

PUBLIC

- ☐ 37 Church
- ☐ 38 Hospital
- ☐ 39 Park/Playground
- ☐ 40 Parking Lot
- ☐ 41 Public Building
- ☐ 42 School
- ☐ 43 Shopping Mall
- ☐ 44 Street/Hwy/Alley
- ☐ 45 Other _____

VEHICLE

- ☐ 46 Camper
- ☐ 47 Motor Home
- ☐ 48 Passenger Car
- ☐ 49 Pick-up
- ☐ 50 Trailer
- ☐ 51 Truck
- ☐ 52 Van
- ☐ 53 Other _____

POINT OF ENTRY 8 — **Q04**

- ☐ 0 Unknown
- ☐ 0 N/A
- ☐ 1 Front
- ☐ 2 Rear
- ☐ 3 Side
- ☐ 4 Door
- ☐ 5 Window
- ☐ 6 Sliding Glass Door
- ☐ 7 Basement
- ☐ 8 Roof
- ☐ 9 Floor
- ☐ 10 Wall
- ☐ 11 Duct/Vent
- ☐ 12 Garage
- ☐ 13 Adj. Building
- ☐ 14 Ground Level
- ☐ 15 Upper Level
- ☐ 16 Other _____

METHOD OF ENTRY 9 — **Q05**

- ☐ 0 Unknown
- ☐ 0 N/A
- ☐ 1 No Force Used
- ☐ 2 Attempt Only
- ☐ 3 Bodily Force
- ☐ 4 Bolt Cut/Pilers
- ☐ 5 Channel Lock/Pipe Wrench/Vice Grips
- ☐ 6 Saw/Drill/Burn
- ☐ 7 Screwdriver
- ☐ 8 Tire Iron
- ☐ 9 Unk Pry Bar
- ☐ 10 Coat Hanger Wire
- ☐ 11 Key Slip Shim
- ☐ 12 Punch
- ☐ 13 Remove Louvers
- ☐ 14 Window Smash
- ☐ 15 Brick/Rock
- ☐ 16 Hid inBuilding
- ☐ 17 Other _____

VEHICLE ENTRY 10 — **Q05**

- ☐ 0 Unknown
- ☐ 0 N/A
- ☐ 1 Door/Lock Forced
- ☐ 2 Trunk Forced
- ☐ 3 Window Broken _____
- ☐ 4 Window Forced _____
- ☐ 5 Window Open _____
- ☐ 6 Unlocked
- ☐ 7 Other _____

PROPERTY ATTACKED 11 — **Q07**

- ☐ 0 Unknown
- ☐ 0 N/A
- ☐ 1 Cash Notes
- ☐ 2 Clothes/Fur
- ☐ 3 Consumable Goods
- ☐ 4 Firearms
- ☐ 5 Household Goods
- ☐ 6 Jewelry Metals
- ☐ 7 Livestock
- ☐ 8 Office Equipment
- ☐ 9 TV/Radio/Camera
- ☐ 10 Miscellaneous
- ☐ 11 Other _____

SEX CRIMES ONLY 12

- ☐ 1 Suspect Climaxed
- ☐ 2 Unknown/Climaxed
- ☐ 3 Victim Bound/Tied
- ☐ 4 Victim Injured
- ☐ 5 Covered Victim Face
- ☐ 6 Photographed Victim
- ☐ 7 Vic Orally Coupulated Susp
- ☐ 8 Susp OrallyCoupulated Vic
- ☐ 9 Rape By Instrument (Foreign Objects)
- ☐ 10 Sodomy
- ☐ 11 Suggested Vic Commit Lewd Perverted Act
- ☐ 12 Inserted Finger into Vagina
- ☐ 13 Forced Vic to Fondle Susp
- ☐ 14 Susp Fondled Victim
- ☐ 15 Masturbated Self
- ☐ 16 Other _____

BURGLARY ONLY 13 — **Q09**

Is member of Neigh Watch?
☐ YES   ☐ NO

Is member of Operation Ident
☐ YES   ☐ NO

Interested in NW?
☐ YES   ☐ NO

Had Home Business Inspection
☐ YES   ☐ NO

When? _____
_____
_____
_____

SUSPECT(S) ACTIONS 14 — **Q10**

- ☐ 1 Alarm Dismarmed
- ☐ 2 Arson
- ☐ 3 Ate/Drank on Premises
- ☐ 4 Blindfolded Victim Bound/Gagged
- ☐ 5 Cat Burglar
- ☐ 6 Defecated/Urinated
- ☐ 7 Demanded Money
- ☐ 8 Disrobed Victim Fully
- ☐ 9 Disrobed Victim Partially
- ☐ 10 Fired Weapon
- ☐ 11 Forced Vic to Move
- ☐ 12 Forced Vic into Veh
- ☐ 13 Has Been Drinking
- ☐ 14 Indication Multi susps.
- ☐ 15 Inflicted injuries
- ☐ 16 Knew Location of Hidden Cash
- ☐ 17 Made Threats
- ☐ 18 Placed Property in Sack/Pocket
- ☐ 19 Prepared Exit
- ☐ 20 Ransacked
- ☐ 21 Ripped/Cut Clothing
- ☐ 22 Selective in Loot
- ☐ 23 Shut Off Power
- ☐ 24 Smoked on Premises
- ☐ 25 Searched Victim
- ☐ 26 Struck Victim
- ☐ 27 Susp Armed
- ☐ 28 Threatened Retaliation
- ☐ 29 Took Only Consumables
- ☐ 30 Took Victim's Vehicle
- ☐ 31 Tortured
- ☐ 32 Under Influence Drugs
- ☐ 33 Used Demand Note
- ☐ 34 Used Lockout
- ☐ 35 Used Driver
- ☐ 36 Used Match/Candle
- ☐ 37 Used Victim Name
- ☐ 38 Used Victim's Suitcase/Pillowcase
- ☐ 39 Used Victim's Tools]
- ☐ 40 Veh Needed to Remove Property
- ☐ 41 Cut/Disconnected Phone
- ☐ 42 Cased Location Before Crime
- ☐ 44 Other _____
_____
_____
_____

SUSP. PRET. TO BE 15 — **Q11**

- ☐ 0 N/A
- ☐ 1 Conducting Survey
- ☐ 2 Cust./Client
- ☐ 3 Delivery Person
- ☐ 4 Disabled Motorist
- ☐ 5 Drunk
- ☐ 6 Employee/Employer
- ☐ 7 Friend/Relative
- ☐ 8 Ill/Injured
- ☐ 9 Need Phone
- ☐ 10 Police/Law
- ☐ 11 Renter
- ☐ 12 Repairman
- ☐ 13 Sale of Illicit Goods
- ☐ 14 Sales Serson
- ☐ 15 Seek Assistance
- ☐ 16 Seek Directions
- ☐ 17 Seeking Someone
- ☐ 18 Solicit Funds
- ☐ 19 Other _____
_____
_____
_____
_____

PHYSICAL SECURITY 16 — **Q12**

- ☐ 0 Unknown
- ☐ 0 N/A
- ☐ 1 Audible Alarm
- ☐ 2 Slient Alarm
- ☐ 3 Private Security Patrol
- ☐ 4 Dog
- ☐ 5 Standard Locks
- ☐ 6 Auxiliary Locks (Deadbolt Windows, etc.)
- ☐ 7 Window Bars/Grills
- ☐ 8 Outside Lighting On
- ☐ 9 Inside Lighting On
- ☐ 10 Garage Door Locked
- ☐ 11 Obscured Interior View (Commercial/Business)
- ☐ 12 Security Signing (N.W., Alarm, etc.)
- ☐ 13 Other _____
_____
_____

VICTIM PROFILE

PHYSICAL CONDITION 17 — **Q13**

- ☐ 0 No Impairment
- ☐ 1 Under Infl. Alcohol/Drugs
- ☐ 2 Sick/Injured
- ☐ 3 Senior Citizen
- ☐ 4 Blind
- ☐ 5 Handicapped
- ☐ 6 Deaf
- ☐ 7 Mute
- ☐ 8 Mentally/Emotionally Impaired
- ☐ 9 Other _____

RELATIONSHIP TO SUSPECT 18 — **Q14**

- ☐ 0 Unknown
- ☐ 1 Husband
- ☐ 2 Wife
- ☐ 3 Mother
- ☐ 4 Father
- ☐ 5 Daughter
- ☐ 6 Son
- ☐ 7 Brother
- ☐ 8 Sister
- ☐ 9 Other Family
- ☐ 10 Acquaintance
- ☐ 11 Friend
- ☐ 12 Boyfriend
- ☐ 13 Girlfriend
- ☐ 14 Neighbor
- ☐ 15 Business Associate
- ☐ 16 Stranger
- ☐ 17 Other _____

MARTIAL STATUS 19 — **Q15**

- ☐ 0 Unkown
- ☐ 1 Annulled
- ☐ 2 Common Law
- ☐ 3 Single
- ☐ 4 Married
- ☐ 5 Divorced
- ☐ 6 Window(er)
- ☐ 7 Separated
- ☐ 8 Other _____

(Continued)

□ NO PROSECUTION DESIRED	**EL SEGUNDO POLICE DEPARTMENT**	A □ ACTIVE	CASE NUMBER
□ TELEPHONE REPORT	348 MAIN STREET	S □ SUSPENDED	
□ INSURANCE REPORT	EL SEGUNDO, CA 90245	R □ RECORDS	
□ COURTESY REPORT	310-322-9114	C □ CLOSED	REFER OTHER RPTS
□ DOMESTIC VIOLENCE	**CRIME REPORT**	K □ COURTESY	
□ CONFIDENTIAL SEX CRIME		U □ UNFOUNDED	

CRIME

CODE SECTION	CRIME			UCR CODE	SECONDARY-COUNTS	OTHER-COUNTS
SPECIFIC LOCATION OF CRIME			OCCURRED ON/OR BETWEEN:	DATE	DAY	TIME
BUSINESS NAME		DATE RPT'D	TIME RPT'D	AND: DATE	DAY	TIME

VICTIM

NAME (Last, First, Middle)	OCCUPATION	D.O.B.	AGE	SEX □ 1. M □ 2. F	RACE □ 1. WHT □ 2. HISP □ 3. BLK □ 5. CHI □ 7. FIL □ 9. P.ISL □ 4. IND □ 6. JAP □ 8. OTH.___
RESIDENCE ADDRESS	CITY	ZIP CODE		RES. PHONE ( )	
BUSINESS NAME AND ADDRESS	CITY	ZIP CODE		BUS. PHONE ( )	

VICTIM(S) - WITNESS - RP

CODE	NAME (Last, First, Middle)	OCCUPATION	D.O.B.	AGE	SEX □ 1. M □ 2. F	RACE □ 1. WHT □ 2. HISP □ 3. BLK □ 5. CHI □ 7. FIL □ 9. P.ISL □ 4. IND □ 6. JAP □ 8. OTH.___
RESIDENCE ADDRESS		CITY	ZIP CODE		RES. PHONE ( )	
BUSINESS NAME AND ADDRESS		CITY	ZIP CODE		BUS. PHONE ( )	
CODE	NAME (Last, First, Middle)	OCCUPATION	D.O.B.	AGE	SEX □ 1. M □ 2. F	RACE □ 1. WHT □ 2. HISP □ 3. BLK □ 5. CHI □ 7. FIL □ 9. P.ISL □ 4. IND □ 6. JAP □ 8. OTH.___
RESIDENCE ADDRESS		CITY	ZIP CODE		RES. PHONE ( )	
BUSINESS NAME AND ADDRESS		CITY	ZIP CODE		BUS. PHONE ( )	
CODE	NAME (Last, First, Middle)	OCCUPATION	D.O.B.	AGE	SEX □ 1. M □ 2. F	RACE □ 1. WHT □ 2. HISP □ 3. BLK □ 5. CHI □ 7. FIL □ 9. P.ISL □ 4. IND □ 6. JAP □ 8. OTH.___
RESIDENCE ADDRESS		CITY	ZIP CODE		RES. PHONE ( )	
BUSINESS ADDRESS		CITY	ZIP CODE		BUS. PHONE ( )	

VIC VEH

LICENSE #	STATE	YEAR	MAKE	MODEL	BODY STYLE □ 0 UNK □ 2 4-DR □ 4 P/U □ 6 VAN □ 8 RV □ 10 OTHER □ 1 2-DR □ 3 CONV □ 5 TRUCK □ 7 S/W □ 9 M/C
COLOR/COLOR		OTHER CHARACTERISTICS (i.e. T/C Damage, Unique Marks or Paint, etc.)			DISPOSITION OF VEHICLE

FACTORS

- □ 1 THERE IS A WITNESS TO THE CRIME   SUSPECT PAGE □ YES □ NO
- □ 2 A SUSPECT WAS ARRESTED
- □ 3 A SUSPECT WAS NAMED
- □ 4 A SUSPECT CAN BE LOCATED
- □ 5 A SUSPECT CAN BE DESCRIBED
- □ 6 A SUSPECT CAN BE IDENTIFIED
- □ 7 A SUSPECT VEHICLE CAN BE IDENTIFIED
- □ 8 THERE IS IDENTIFIABLE STOLEN PROPERTY
- □ 9 THERE IS A SIGNIFIANT M.O.
- □ 10 SIGNIFICANT PHYSICAL EVIDENCE IS PRESENT
- □ 11 THERE IS A MAJOR INJURY/SEX CRIME INVOLVED
- □ 12 THERE IS A GOOD POSSIBILITY OF A SOLUTION
- □ 13 FURTHER INVESTIGATION NEEDED
- □ 14 CRIME IS GANG RELATED
- □ 15 HATE CRIME RELATED

EVIDENCE

- □ 0 NONE
- □ 1 FINGERPRINTS
- □ 2 TOOLS
- □ 3 TOOL MARKINGS
- □ 4 GLASS
- □ 5 PAINT
- □ 6 BULLET CASING
- □ 7 BULLET
- □ 8 RAPE KIT
- □ 9 SEMEN
- □ 10 BLOOD
- □ 11 URINE
- □ 12 HAIR
- □ 13 FIREARMS
- □ 14 PHOTOGRAPHS
- □ 15 OTHER (DESCRIBE)

VICTIMS SIGNATURE	DATE	DETECTIVE ASSIGNED SIGNATURE	DATE

REPORTING OFFICER	ID#	DATE	REVIEWING SUPERVISOR	ID#	DATE

COPIES: TO:	□ CHIEF □ CII □ PATROL □ DMV □ CAU □ ABC (2 copies)	□ DB □ DA	□ OTHER	ROUTED BY	ENTERED BY

ESPD Form 3/97

El Segundo Police Department Crime Report (Front and Reverse)

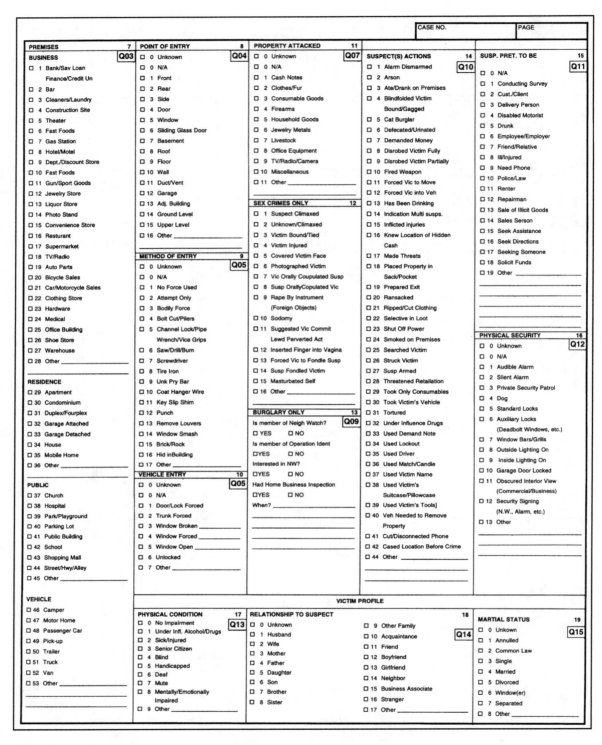

CASE NO.		PAGE	

PREMISES   7   Q03

BUSINESS
- ☐ 1 Bank/Sav Loan Finance/Credit Un
- ☐ 2 Bar
- ☐ 3 Cleaners/Laundry
- ☐ 4 Construction Site
- ☐ 5 Theater
- ☐ 6 Fast Foods
- ☐ 7 Gas Station
- ☐ 8 Hotel/Motel
- ☐ 9 Dept./Discount Store
- ☐ 10 Fast Foods
- ☐ 11 Gun/Sport Goods
- ☐ 12 Jewelry Store
- ☐ 13 Liquor Store
- ☐ 14 Photo Stand
- ☐ 15 Convenience Store
- ☐ 16 Resturant
- ☐ 17 Supermarket
- ☐ 18 TV/Radio
- ☐ 19 Auto Parts
- ☐ 20 Bicycle Sales
- ☐ 21 Car/Motorcycle Sales
- ☐ 22 Clothing Store
- ☐ 23 Hardware
- ☐ 24 Medical
- ☐ 25 Office Building
- ☐ 26 Shoe Store
- ☐ 27 Warehouse
- ☐ 28 Other _____

RESIDENCE
- ☐ 29 Apartment
- ☐ 30 Condominium
- ☐ 31 Duplex/Fourplex
- ☐ 32 Garage Attached
- ☐ 33 Garage Detached
- ☐ 34 House
- ☐ 35 Mobile Home
- ☐ 36 Other _____

PUBLIC
- ☐ 37 Church
- ☐ 38 Hospital
- ☐ 39 Park/Playground
- ☐ 40 Parking Lot
- ☐ 41 Public Building
- ☐ 42 School
- ☐ 43 Shopping Mall
- ☐ 44 Street/Hwy/Alley
- ☐ 45 Other _____

VEHICLE
- ☐ 46 Camper
- ☐ 47 Motor Home
- ☐ 48 Passenger Car
- ☐ 49 Pick-up
- ☐ 50 Trailer
- ☐ 51 Truck
- ☐ 52 Van
- ☐ 53 Other _____

POINT OF ENTRY   8   Q04
- ☐ 0 Unknown
- ☐ 0 N/A
- ☐ 1 Front
- ☐ 2 Rear
- ☐ 3 Side
- ☐ 4 Door
- ☐ 5 Window
- ☐ 6 Sliding Glass Door
- ☐ 7 Basement
- ☐ 8 Roof
- ☐ 9 Floor
- ☐ 10 Wall
- ☐ 11 Duct/Vent
- ☐ 12 Garage
- ☐ 13 Adj. Building
- ☐ 14 Ground Level
- ☐ 15 Upper Level
- ☐ 16 Other _____

METHOD OF ENTRY   9   Q05
- ☐ 0 Unknown
- ☐ 0 N/A
- ☐ 1 No Force Used
- ☐ 2 Attempt Only
- ☐ 3 Bodily Force
- ☐ 4 Bolt Cut/Pliers
- ☐ 5 Channel Lock/Pipe Wrench/Vice Grips
- ☐ 6 Saw/Drill/Burn
- ☐ 7 Screwdriver
- ☐ 8 Tire Iron
- ☐ 9 Unk Pry Bar
- ☐ 10 Coat Hanger Wire
- ☐ 11 Key Slip Shim
- ☐ 12 Punch
- ☐ 13 Remove Louvers
- ☐ 14 Window Smash
- ☐ 15 Brick/Rock
- ☐ 16 Hid inBuilding
- ☐ 17 Other _____

VEHICLE ENTRY   10   Q05
- ☐ 0 Unknown
- ☐ 0 N/A
- ☐ 1 Door/Lock Forced
- ☐ 2 Trunk Forced
- ☐ 3 Window Broken _____
- ☐ 4 Window Forced _____
- ☐ 5 Window Open _____
- ☐ 6 Unlocked
- ☐ 7 Other _____

PROPERTY ATTACKED   11   Q07
- ☐ 0 Unknown
- ☐ 0 N/A
- ☐ 1 Cash Notes
- ☐ 2 Clothes/Fur
- ☐ 3 Consumable Goods
- ☐ 4 Firearms
- ☐ 5 Household Goods
- ☐ 6 Jewelry Metals
- ☐ 7 Livestock
- ☐ 8 Office Equipment
- ☐ 9 TV/Radio/Camera
- ☐ 10 Miscellaneous
- ☐ 11 Other _____

SEX CRIMES ONLY   12
- ☐ 1 Suspect Climaxed
- ☐ 2 Unknown/Climaxed
- ☐ 3 Victim Bound/Tied
- ☐ 4 Victim Injured
- ☐ 5 Covered Victim Face
- ☐ 6 Photographed Victim
- ☐ 7 Vic Orally Copulated Susp
- ☐ 8 Susp OrallyCopulated Vic
- ☐ 9 Rape By Instrument (Foreign Objects)
- ☐ 10 Sodomy
- ☐ 11 Suggested Vic Commit Lewd Perverted Act
- ☐ 12 Inserted Finger into Vagina
- ☐ 13 Forced Vic to Fondle Susp
- ☐ 14 Susp Fondled Victim
- ☐ 15 Masturbated Self
- ☐ 16 Other _____

BURGLARY ONLY   13   Q09
Is member of Neigh Watch?
☐ YES   ☐ NO
Is member of Operation Ident
☐ YES   ☐ NO
Interested in NW?
☐ YES   ☐ NO
Had Home Business Inspection
☐ YES   ☐ NO
When? _____
_____
_____
_____

SUSPECT(S) ACTIONS   14   Q10
- ☐ 1 Alarm Dismarmed
- ☐ 2 Arson
- ☐ 3 Ate/Drank on Premises
- ☐ 4 Blindfolded Victim Bound/Gagged
- ☐ 5 Cat Burglar
- ☐ 6 Defecated/Urinated
- ☐ 7 Demanded Money
- ☐ 8 Disrobed Victim Fully
- ☐ 9 Disrobed Victim Partially
- ☐ 10 Fired Weapon
- ☐ 11 Forced Vic to Move
- ☐ 12 Forced Vic into Veh
- ☐ 13 Has Been Drinking
- ☐ 14 Indication Multi susps.
- ☐ 15 Inflicted injuries
- ☐ 16 Knew Location of Hidden Cash
- ☐ 17 Made Threats
- ☐ 18 Placed Property in Sack/Pocket
- ☐ 19 Prepared Exit
- ☐ 20 Ransacked
- ☐ 21 Ripped/Cut Clothing
- ☐ 22 Selective in Loot
- ☐ 23 Shut Off Power
- ☐ 24 Smoked on Premises
- ☐ 25 Searched Victim
- ☐ 26 Struck Victim
- ☐ 27 Susp Armed
- ☐ 28 Threatened Retaliation
- ☐ 29 Took Only Consumables
- ☐ 30 Took Victim's Vehicle
- ☐ 31 Tortured
- ☐ 32 Under Influence Drugs
- ☐ 33 Used Demand Note
- ☐ 34 Used Lockout
- ☐ 35 Used Driver
- ☐ 36 Used Match/Candle
- ☐ 37 Used Victim Name
- ☐ 38 Used Victim's Suitcase/Pillowcase
- ☐ 39 Used Victim's Tools]
- ☐ 40 Veh Needed to Remove Property
- ☐ 41 Cut/Disconnected Phone
- ☐ 42 Cased Location Before Crime
- ☐ 44 Other _____
_____
_____

SUSP. PRET. TO BE   15   Q11
- ☐ 0 N/A
- ☐ 1 Conducting Survey
- ☐ 2 Cust./Client
- ☐ 3 Delivery Person
- ☐ 4 Disabled Motorist
- ☐ 5 Drunk
- ☐ 6 Employee/Employer
- ☐ 7 Friend/Relative
- ☐ 8 Ill/Injured
- ☐ 9 Need Phone
- ☐ 10 Police/Law
- ☐ 11 Renter
- ☐ 12 Repairman
- ☐ 13 Sale of Illicit Goods
- ☐ 14 Sales Serson
- ☐ 15 Seek Assistance
- ☐ 16 Seek Directions
- ☐ 17 Seeking Someone
- ☐ 18 Solicit Funds
- ☐ 19 Other _____
_____
_____
_____
_____

PHYSICAL SECURITY   16   Q12
- ☐ 0 Unknown
- ☐ 0 N/A
- ☐ 1 Audible Alarm
- ☐ 2 Silent Alarm
- ☐ 3 Private Security Patrol
- ☐ 4 Dog
- ☐ 5 Standard Locks
- ☐ 6 Auxiliary Locks (Deadbolt Windows, etc.)
- ☐ 7 Window Bars/Grills
- ☐ 8 Outside Lighting On
- ☐ 9 Inside Lighting On
- ☐ 10 Garage Door Locked
- ☐ 11 Obscured Interior View (Commercial/Business)
- ☐ 12 Security Signing (N.W., Alarm, etc.)
- ☐ 13 Other _____
_____
_____

VICTIM PROFILE

PHYSICAL CONDITION   17   Q13
- ☐ 0 No Impairment
- ☐ 1 Under Infl. Alcohol/Drugs
- ☐ 2 Sick/Injured
- ☐ 3 Senior Citizen
- ☐ 4 Blind
- ☐ 5 Handicapped
- ☐ 6 Deaf
- ☐ 7 Mute
- ☐ 8 Mentally/Emotionally Impaired
- ☐ 9 Other _____

RELATIONSHIP TO SUSPECT   18   Q14
- ☐ 0 Unknown
- ☐ 1 Husband
- ☐ 2 Wife
- ☐ 3 Mother
- ☐ 4 Father
- ☐ 5 Daughter
- ☐ 6 Son
- ☐ 7 Brother
- ☐ 8 Sister
- ☐ 9 Other Family
- ☐ 10 Acquaintance
- ☐ 11 Friend
- ☐ 12 Boyfriend
- ☐ 13 Girlfriend
- ☐ 14 Neighbor
- ☐ 15 Business Associate
- ☐ 16 Stranger
- ☐ 17 Other _____

MARTIAL STATUS   19   Q15
- ☐ 0 Unknown
- ☐ 1 Annulled
- ☐ 2 Common Law
- ☐ 3 Single
- ☐ 4 Married
- ☐ 5 Divorced
- ☐ 6 Window(er)
- ☐ 7 Separated
- ☐ 8 Other _____

(Continued)

| ☐ NO PROSECUTION DESIRED
☐ TELEPHONE REPORT
☐ INSURANCE REPORT
☐ COURTESY REPORT
☐ DOMESTIC VIOLENCE
☐ CONFIDENTIAL SEX CRIME | **EL SEGUNDO POLICE DEPARTMENT**
348 MAIN STREET
EL SEGUNDO, CA 90245
310-322-9114

CRIME REPORT | A ☐ ACTIVE
S ☐ SUSPENDED
R ☐ RECORDS
C ☐ CLOSED
K ☐ COURTESY
U ☐ UNFOUNDED | CASE NUMBER

REFER OTHER RPTS |

CRIME

| CODE SECTION | CRIME | | UCR CODE | SECONDARY-COUNTS | OTHER-COUNTS |

| SPECIFIC LOCATION OF CRIME | | OCCURRED ON/OR BETWEEN: | DATE | DAY | TIME |

| BUSINESS NAME | DATE RPT'D | TIME RPT'D | AND: | DATE | DAY | TIME |

VICTIM

| NAME (Last, First, Middle) | OCCUPATION | D.O.B. | AGE | SEX
☐ 1. M
☐ 2. F | RACE ☐ 1. WHT ☐ 2. HISP
☐ 3. BLK ☐ 5. CHI ☐ 7. FIL ☐ 9. P.ISL.
☐ 4. IND ☐ 6. JAP ☐ 8. OTH.____ |

| RESIDENCE ADDRESS | CITY | ZIP CODE | RES. PHONE
(   ) |

| BUSINESS NAME AND ADDRESS | CITY | ZIP CODE | BUS. PHONE
(   ) |

VICTIM(S) - WITNESS - RP

| CODE | NAME (Last, First, Middle) | OCCUPATION | D.O.B. | AGE | SEX
☐ 1. M
☐ 2. F | RACE ☐ 1. WHT ☐ 2. HISP
☐ 3. BLK ☐ 5. CHI ☐ 7. FIL ☐ 9. P.ISL.
☐ 4. IND ☐ 6. JAP ☐ 8. OTH.____ |

| RESIDENCE ADDRESS | CITY | ZIP CODE | RES. PHONE
(   ) |

| BUSINESS NAME AND ADDRESS | CITY | ZIP CODE | BUS. PHONE
(   ) |

| CODE | NAME (Last, First, Middle) | OCCUPATION | D.O.B. | AGE | SEX
☐ 1. M
☐ 2. F | RACE ☐ 1. WHT ☐ 2. HISP
☐ 3. BLK ☐ 5. CHI ☐ 7. FIL ☐ 9. P.ISL.
☐ 4. IND ☐ 6. JAP ☐ 8. OTH.____ |

| RESIDENCE ADDRESS | CITY | ZIP CODE | RES. PHONE
(   ) |

| BUSINESS NAME AND ADDRESS | CITY | ZIP CODE | BUS. PHONE
(   ) |

| CODE | NAME (Last, First, Middle) | OCCUPATION | D.O.B. | AGE | SEX
☐ 1. M
☐ 2. F | RACE ☐ 1. WHT ☐ 2. HISP
☐ 3. BLK ☐ 5. CHI ☐ 7. FIL ☐ 9. P.ISL.
☐ 4. IND ☐ 6. JAP ☐ 8. OTH.____ |

| RESIDENCE ADDRESS | CITY | ZIP CODE | RES. PHONE
(   ) |

| BUSINESS ADDRESS | CITY | ZIP CODE | BUS. PHONE
(   ) |

VIC VEH

| LICENSE # | STATE | YEAR | MAKE | MODEL | BODY STYLE ☐ 0 UNK ☐ 2 4-DR ☐ 4 P/U ☐ 6 VAN ☐ 8 RV ☐ 10 OTHER
☐ 1 2-DR ☐ 3 CONV ☐ 5 TRUCK ☐ 7 S/W ☐ 9 M/C |

| COLOR/COLOR | OTHER CHARACTERISTICS (i.e. T/C Damage, Unique Marks or Paint, etc.) | DISPOSITION OF VEHICLE |

FACTORS

☐ 1 THERE IS A WITNESS TO THE CRIME   SUSPECT PAGE ☐ YES ☐ NO
☐ 2 A SUSPECT WAS ARRESTED
☐ 3 A SUSPECT WAS NAMED
☐ 4 A SUSPECT CAN BE LOCATED
☐ 5 A SUSPECT CAN BE DESCRIBED
☐ 6 A SUSPECT CAN BE IDENTIFIED
☐ 7 A SUSPECT VEHICLE CAN BE IDENTIFIED
☐ 8 THERE IS IDENTIFIABLE STOLEN PROPERTY
☐ 9 THERE IS A SIGNIFIANT M.O.
☐ 10 SIGNIFICANT PHYSICAL EVIDENCE IS PRESENT
☐ 11 THERE IS A MAJOR INJURY/SEX CRIME INVOLVED
☐ 12 THERE IS A GOOD POSSIBILITY OF A SOLUTION
☐ 13 FURTHER INVESTIGATION NEEDED
☐ 14 CRIME IS GANG RELATED
☐ 15 HATE CRIME RELATED

EVIDENCE

☐ 0 NONE   ☐ 10 BLOOD
☐ 1 FINGERPRINTS   ☐ 11 URINE
☐ 2 TOOLS   ☐ 12 HAIR
☐ 3 TOOL MARKINGS   ☐ 13 FIREARMS
☐ 4 GLASS   ☐ 14 PHOTOGRAPHS
☐ 5 PAINT   ☐ 15 OTHER (DESCRIBE)
☐ 6 BULLET CASING
☐ 7 BULLET
☐ 8 RAPE KIT
☐ 9 SEMEN

| VICTIMS SIGNATURE | DATE | DETECTIVE ASSIGNED SIGNATURE | DATE |

| REPORTING OFFICER | ID# | DATE | REVIEWING SUPERVISOR | ID# | DATE |

| COPIES: ☐ CHIEF ☐ CII ☐ PATROL ☐ DB ☐ OTHER
TO: ☐ DMV ☐ CAU ☐ ABC (2 copies) ☐ DA ____ | ROUTED BY | ENTERED BY |

ESPD Form 3/97

El Segundo Police Department Crime Report (Front and Reverse)

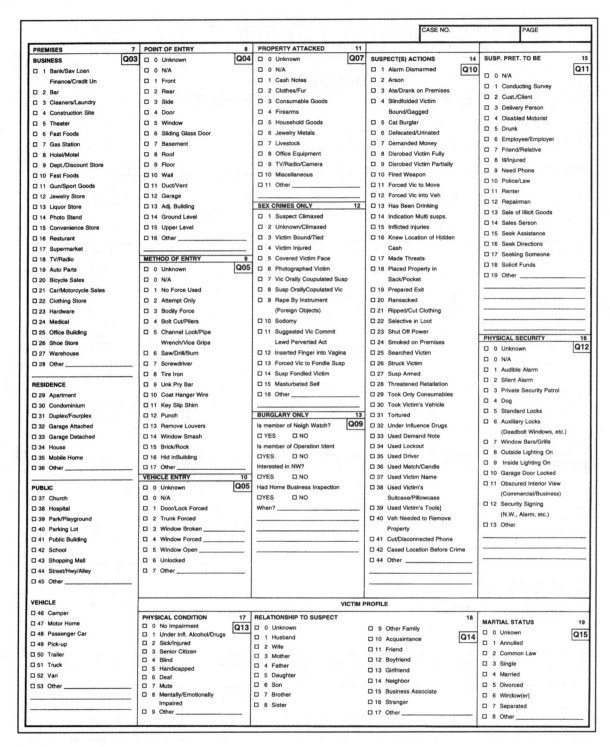

CASE NO.		PAGE

PREMISES 7

BUSINESS Q03
- □ 1 Bank/Sav Loan Finance/Credit Un
- □ 2 Bar
- □ 3 Cleaners/Laundry
- □ 4 Construction Site
- □ 5 Theater
- □ 6 Fast Foods
- □ 7 Gas Station
- □ 8 Hotel/Motel
- □ 9 Dept./Discount Store
- □ 10 Fast Foods
- □ 11 Gun/Sport Goods
- □ 12 Jewelry Store
- □ 13 Liquor Store
- □ 14 Photo Stand
- □ 15 Convenience Store
- □ 16 Resturant
- □ 17 Supermarket
- □ 18 TV/Radio
- □ 19 Auto Parts
- □ 20 Bicycle Sales
- □ 21 Car/Motorcycle Sales
- □ 22 Clothing Store
- □ 23 Hardware
- □ 24 Medical
- □ 25 Office Building
- □ 26 Shoe Store
- □ 27 Warehouse
- □ 28 Other _____

RESIDENCE
- □ 29 Apartment
- □ 30 Condominium
- □ 31 Duplex/Fourplex
- □ 32 Garage Attached
- □ 33 Garage Detached
- □ 34 House
- □ 35 Mobile Home
- □ 36 Other _____

PUBLIC
- □ 37 Church
- □ 38 Hospital
- □ 39 Park/Playground
- □ 40 Parking Lot
- □ 41 Public Building
- □ 42 School
- □ 43 Shopping Mall
- □ 44 Street/Hwy/Alley
- □ 45 Other _____

VEHICLE
- □ 46 Camper
- □ 47 Motor Home
- □ 48 Passenger Car
- □ 49 Pick-up
- □ 50 Trailer
- □ 51 Truck
- □ 52 Van
- □ 53 Other _____

POINT OF ENTRY 8 Q04
- □ 0 Unknown
- □ 0 N/A
- □ 1 Front
- □ 2 Rear
- □ 3 Side
- □ 4 Door
- □ 5 Window
- □ 6 Sliding Glass Door
- □ 7 Basement
- □ 8 Roof
- □ 9 Floor
- □ 10 Wall
- □ 11 Duct/Vent
- □ 12 Garage
- □ 13 Adj. Building
- □ 14 Ground Level
- □ 15 Upper Level
- □ 16 Other _____

METHOD OF ENTRY 9 Q05
- □ 0 Unknown
- □ 0 N/A
- □ 1 No Force Used
- □ 2 Attempt Only
- □ 3 Bodily Force
- □ 4 Bolt Cut/Pliers
- □ 5 Channel Lock/Pipe Wrench/Vice Grips
- □ 6 Saw/Drill/Burn
- □ 7 Screwdriver
- □ 8 Tire Iron
- □ 9 Unk Pry Bar
- □ 10 Coat Hanger Wire
- □ 11 Key Slip Shim
- □ 12 Punch
- □ 13 Remove Louvers
- □ 14 Window Smash
- □ 15 Brick/Rock
- □ 16 Hid inBuilding
- □ 17 Other _____

VEHICLE ENTRY 10 Q05
- □ 0 Unknown
- □ 0 N/A
- □ 1 Door/Lock Forced
- □ 2 Trunk Forced
- □ 3 Window Broken _____
- □ 4 Window Forced
- □ 5 Window Open _____
- □ 6 Unlocked
- □ 7 Other _____

PROPERTY ATTACKED 11 Q07
- □ 0 Unknown
- □ 0 N/A
- □ 1 Cash Notes
- □ 2 Clothes/Fur
- □ 3 Consumable Goods
- □ 4 Firearms
- □ 5 Household Goods
- □ 6 Jewelry Metals
- □ 7 Livestock
- □ 8 Office Equipment
- □ 9 TV/Radio/Camera
- □ 10 Miscellaneous
- □ 11 Other _____

SEX CRIMES ONLY 12
- □ 1 Suspect Climaxed
- □ 2 Unknown/Climaxed
- □ 3 Victim Bound/Tied
- □ 4 Victim Injured
- □ 5 Covered Victim Face
- □ 6 Photographed Victim
- □ 7 Vic Orally Coupulated Susp
- □ 8 Susp OrallyCopulated Vic
- □ 9 Rape By Instrument (Foreign Objects)
- □ 10 Sodomy
- □ 11 Suggested Vic Commit Lewd Perverted Act
- □ 12 Inserted Finger into Vagina
- □ 13 Forced Vic to Fondle Susp
- □ 14 Susp Fondled Victim
- □ 15 Masturbated Self
- □ 16 Other _____

BURGLARY ONLY 13 Q09
- Is member of Neigh Watch?
- □ YES □ NO
- Is member of Operation Ident
- □ YES □ NO
- Interested in NW?
- □ YES □ NO
- Had Home Business Inspection
- □ YES □ NO
- When? _____
- _____
- _____

VICTIM PROFILE

PHYSICAL CONDITION 17 Q13
- □ 0 No Impairment
- □ 1 Under Infl. Alcohol/Drugs
- □ 2 Sick/Injured
- □ 3 Senior Citizen
- □ 4 Blind
- □ 5 Handicapped
- □ 6 Deaf
- □ 7 Mute
- □ 8 Mentally/Emotionally Impaired
- □ 9 Other _____

SUSPECT(S) ACTIONS 14
- □ 1 Alarm Dismarmed Q10
- □ 2 Arson
- □ 3 Ate/Drank on Premises
- □ 4 Blindfolded Victim Bound/Gagged
- □ 5 Cat Burglar
- □ 6 Defecated/Urinated
- □ 7 Demanded Money
- □ 8 Disrobed Victim Fully
- □ 9 Disrobed Victim Partially
- □ 10 Fired Weapon
- □ 11 Forced Vic to Move
- □ 12 Forced Vic into Veh
- □ 13 Has Been Drinking
- □ 14 Indication Multi susps.
- □ 15 Inflicted injuries
- □ 16 Knew Location of Hidden Cash
- □ 17 Made Threats
- □ 18 Placed Property in Sack/Pocket
- □ 19 Prepared Exit
- □ 20 Ransacked
- □ 21 Ripped/Cut Clothing
- □ 22 Selective in Loot
- □ 23 Shut Off Power
- □ 24 Smoked on Premises
- □ 25 Searched Victim
- □ 26 Struck Victim
- □ 27 Susp Armed
- □ 28 Threatened Retaliation
- □ 29 Took Only Consumables
- □ 30 Took Victim's Vehicle
- □ 31 Tortured
- □ 32 Under Influence Drugs
- □ 33 Used Demand Note
- □ 34 Used Lockout
- □ 35 Used Driver
- □ 36 Used Match/Candle
- □ 37 Used Victim Name
- □ 38 Used Victim's Suitcase/Pillowcase
- □ 39 Used Victim's Tools]
- □ 40 Veh Needed to Remove Property
- □ 41 Cut/Disconnected Phone
- □ 42 Cased Location Before Crime
- □ 44 Other _____
- _____
- _____

RELATIONSHIP TO SUSPECT 18 Q14
- □ 0 Unknown
- □ 1 Husband
- □ 2 Wife
- □ 3 Mother
- □ 4 Father
- □ 5 Daughter
- □ 6 Son
- □ 7 Brother
- □ 8 Sister
- □ 9 Other Family
- □ 10 Acquaintance
- □ 11 Friend
- □ 12 Boyfriend
- □ 13 Girlfriend
- □ 14 Neighbor
- □ 15 Business Associate
- □ 16 Stranger
- □ 17 Other _____

SUSP. PRET. TO BE 15 Q11
- □ 0 N/A
- □ 1 Conducting Survey
- □ 2 Cust./Client
- □ 3 Delivery Person
- □ 4 Disabled Motorist
- □ 5 Drunk
- □ 6 Employee/Employer
- □ 7 Friend/Relative
- □ 8 Ill/Injured
- □ 9 Need Phone
- □ 10 Police/Law
- □ 11 Renter
- □ 12 Repairman
- □ 13 Sale of Illicit Goods
- □ 14 Sales Serson
- □ 15 Seek Assistance
- □ 16 Seek Directions
- □ 17 Seeking Someone
- □ 18 Solicit Funds
- □ 19 Other _____
- _____
- _____
- _____

PHYSICAL SECURITY 16 Q12
- □ 0 Unknown
- □ 0 N/A
- □ 1 Audible Alarm
- □ 2 Silent Alarm
- □ 3 Private Security Patrol
- □ 4 Dog
- □ 5 Standard Locks
- □ 6 Auxiliary Locks (Deadbolt Windows, etc.)
- □ 7 Window Bars/Grills
- □ 8 Outside Lighting On
- □ 9 Inside Lighting On
- □ 10 Garage Door Locked
- □ 11 Obscured Interior View (Commercial/Business)
- □ 12 Security Signing (N.W., Alarm, etc.)
- □ 13 Other _____
- _____
- _____

MARTIAL STATUS 19 Q15
- □ 0 Unknown
- □ 1 Annulled
- □ 2 Common Law
- □ 3 Single
- □ 4 Married
- □ 5 Divorced
- □ 6 Window(er)
- □ 7 Separated
- □ 8 Other _____

(*Continued*)

EL SEGUNDO POLICE DEPARTMENT

ADDITIONAL VICTIMS/WITNESSES

PAGE_____ OF _____

CASE NO

CRIME 1	CODE SECTION		CRIME		CLASSIFICATION				REFER OTHER REPORTS	
	LOCATION				RD.	DATE	TIME	SUPPL. ☐	INCIDENT NO.	

2 — WITNESS/RP/OR ADD. VICTIM

CODE	NAME (Last, First, Middle)	OCCUPATION	D.O.B.	AGE	SEX ☐1. M ☐2. F	RACE ☐1. WHT ☐2. HISP ☐3. BLK ☐5. CHI ☐7. FIL ☐9. P.ISL. ☐4. IND ☐6. JAP ☐8. OTH.____
RESIDENCE ADDRESS		CITY	ZIP CODE		RES. PHONE ( )	
BUSINESS ADDRESS		CITY	ZIP CODE		BUS. PHONE ( )	
CODE	NAME (Last, First, Middle)	OCCUPATION	D.O.B.	AGE	SEX ☐1. M ☐2. F	RACE ☐1. WHT ☐2. HISP ☐3. BLK ☐5. CHI ☐7. FIL ☐9. P.ISL. ☐4. IND ☐6. JAP ☐8. OTH.____
RESIDENCE ADDRESS		CITY	ZIP CODE		RES. PHONE ( )	
BUSINESS ADDRESS		CITY	ZIP CODE		BUS. PHONE ( )	
CODE	NAME (Last, First, Middle)	OCCUPATION	D.O.B.	AGE	SEX ☐1. M ☐2. F	RACE ☐1. WHT ☐2. HISP ☐3. BLK ☐5. CHI ☐7. FIL ☐9. P.ISL. ☐4. IND ☐6. JAP ☐8. OTH.____
RESIDENCE ADDRESS		CITY	ZIP CODE		RES. PHONE ( )	
BUSINESS ADDRESS		CITY	ZIP CODE		BUS. PHONE ( )	
CODE	NAME (Last, First, Middle)	OCCUPATION	D.O.B.	AGE	SEX ☐1. M ☐2. F	RACE ☐1. WHT ☐2. HISP ☐3. BLK ☐5. CHI ☐7. FIL ☐9. P.ISL. ☐4. IND ☐6. JAP ☐8. OTH.____
RESIDENCE ADDRESS		CITY	ZIP CODE		RES. PHONE ( )	
BUSINESS ADDRESS		CITY	ZIP CODE		BUS. PHONE ( )	
CODE	NAME (Last, First, Middle)	OCCUPATION	D.O.B.	AGE	SEX ☐1. M ☐2. F	RACE ☐1. WHT ☐2. HISP ☐3. BLK ☐5. CHI ☐7. FIL ☐9. P.ISL. ☐4. IND ☐6. JAP ☐8. OTH.____
RESIDENCE ADDRESS		CITY	ZIP CODE		RES. PHONE ( )	
BUSINESS ADDRESS		CITY	ZIP CODE		BUS. PHONE ( )	
CODE	NAME (Last, First, Middle)	OCCUPATION	D.O.B.	AGE	SEX ☐1. M ☐2. F	RACE ☐1. WHT ☐2. HISP ☐3. BLK ☐5. CHI ☐7. FIL ☐9. P.ISL. ☐4. IND ☐6. JAP ☐8. OTH.____
RESIDENCE ADDRESS		CITY	ZIP CODE		RES. PHONE ( )	
BUSINESS ADDRESS		CITY	ZIP CODE		BUS. PHONE ( )	
CODE	NAME (Last, First, Middle)	OCCUPATION	D.O.B.	AGE	SEX ☐1. M ☐2. F	RACE ☐1. WHT ☐2. HISP ☐3. BLK ☐5. CHI ☐7. FIL ☐9. P.ISL. ☐4. IND ☐6. JAP ☐8. OTH.____
RESIDENCE ADDRESS		CITY	ZIP CODE		RES. PHONE ( )	
BUSINESS ADDRESS		CITY	ZIP CODE		BUS. PHONE ( )	
CODE	NAME (Last, First, Middle)	OCCUPATION	D.O.B.	AGE	SEX ☐1. M ☐2. F	RACE ☐1. WHT ☐2. HISP ☐3. BLK ☐5. CHI ☐7. FIL ☐9. P.ISL. ☐4. IND ☐6. JAP ☐8. OTH.____
RESIDENCE ADDRESS		CITY	ZIP CODE		RES. PHONE ( )	
BUSINESS ADDRESS		CITY	ZIP CODE		BUS. PHONE ( )	

REPORTING OFFICER	ID#	DATE	REVIEWED BY	ID#	DATE

COPIES: ☐ CHIEF ☐ CII ☐ PATROL ☐ DB ☐ OTHER AGENCY — ROUTED BY — ENTERED BY
TO: ☐ DMV ☐ CAU ☐ ABC (2 copies) ☐ DA _____

ESPD Form #269 (Rev 5/97)

El Segundo Police Department Additional Victims/Witnesses

EL SEGUNDO POLICE DEPARTMENT — ADDITIONAL VICTIMS/WITNESSES

PAGE_____ OF _____

CASE NO

CRIME 1

| CODE SECTION | CRIME | | CLASSIFICATION | | | | REFER OTHER REPORTS |

| LOCATION | | | RD. | DATE | TIME | SUPPL. ☐ | INCIDENT NO. |

WITNESS/RP/OR ADD. VICTIM 2

CODE	NAME (Last, First, Middle)	OCCUPATION	D.O.B.	AGE	SEX	RACE
					☐ 1. M ☐ 2. F	☐ 1. WHT ☐ 2. HISP ☐ 3. BLK ☐ 5. CHI ☐ 7. FIL ☐ 9. P.ISL. ☐ 4. IND ☐ 6. JAP ☐ 8. OTH.____

| RESIDENCE ADDRESS | CITY | ZIP CODE | RES. PHONE ( ) |
| BUSINESS ADDRESS | CITY | ZIP CODE | BUS. PHONE ( ) |

CODE	NAME (Last, First, Middle)	OCCUPATION	D.O.B.	AGE	SEX	RACE
					☐ 1. M ☐ 2. F	☐ 1. WHT ☐ 2. HISP ☐ 3. BLK ☐ 5. CHI ☐ 7. FIL ☐ 9. P.ISL. ☐ 4. IND ☐ 6. JAP ☐ 8. OTH.____

| RESIDENCE ADDRESS | CITY | ZIP CODE | RES. PHONE ( ) |
| BUSINESS ADDRESS | CITY | ZIP CODE | BUS. PHONE ( ) |

CODE	NAME (Last, First, Middle)	OCCUPATION	D.O.B.	AGE	SEX	RACE
					☐ 1. M ☐ 2. F	☐ 1. WHT ☐ 2. HISP ☐ 3. BLK ☐ 5. CHI ☐ 7. FIL ☐ 9. P.ISL. ☐ 4. IND ☐ 6. JAP ☐ 8. OTH.____

| RESIDENCE ADDRESS | CITY | ZIP CODE | RES. PHONE ( ) |
| BUSINESS ADDRESS | CITY | ZIP CODE | BUS. PHONE ( ) |

CODE	NAME (Last, First, Middle)	OCCUPATION	D.O.B.	AGE	SEX	RACE
					☐ 1. M ☐ 2. F	☐ 1. WHT ☐ 2. HISP ☐ 3. BLK ☐ 5. CHI ☐ 7. FIL ☐ 9. P.ISL. ☐ 4. IND ☐ 6. JAP ☐ 8. OTH.____

| RESIDENCE ADDRESS | CITY | ZIP CODE | RES. PHONE ( ) |
| BUSINESS ADDRESS | CITY | ZIP CODE | BUS. PHONE ( ) |

CODE	NAME (Last, First, Middle)	OCCUPATION	D.O.B.	AGE	SEX	RACE
					☐ 1. M ☐ 2. F	☐ 1. WHT ☐ 2. HISP ☐ 3. BLK ☐ 5. CHI ☐ 7. FIL ☐ 9. P.ISL. ☐ 4. IND ☐ 6. JAP ☐ 8. OTH.____

| RESIDENCE ADDRESS | CITY | ZIP CODE | RES. PHONE ( ) |
| BUSINESS ADDRESS | CITY | ZIP CODE | BUS. PHONE ( ) |

CODE	NAME (Last, First, Middle)	OCCUPATION	D.O.B.	AGE	SEX	RACE
					☐ 1. M ☐ 2. F	☐ 1. WHT ☐ 2. HISP ☐ 3. BLK ☐ 5. CHI ☐ 7. FIL ☐ 9. P.ISL. ☐ 4. IND ☐ 6. JAP ☐ 8. OTH.____

| RESIDENCE ADDRESS | CITY | ZIP CODE | RES. PHONE ( ) |
| BUSINESS ADDRESS | CITY | ZIP CODE | BUS. PHONE ( ) |

CODE	NAME (Last, First, Middle)	OCCUPATION	D.O.B.	AGE	SEX	RACE
					☐ 1. M ☐ 2. F	☐ 1. WHT ☐ 2. HISP ☐ 3. BLK ☐ 5. CHI ☐ 7. FIL ☐ 9. P.ISL. ☐ 4. IND ☐ 6. JAP ☐ 8. OTH.____

| RESIDENCE ADDRESS | CITY | ZIP CODE | RES. PHONE ( ) |
| BUSINESS ADDRESS | CITY | ZIP CODE | BUS. PHONE ( ) |

CODE	NAME (Last, First, Middle)	OCCUPATION	D.O.B.	AGE	SEX	RACE
					☐ 1. M ☐ 2. F	☐ 1. WHT ☐ 2. HISP ☐ 3. BLK ☐ 5. CHI ☐ 7. FIL ☐ 9. P.ISL. ☐ 4. IND ☐ 6. JAP ☐ 8. OTH.____

| RESIDENCE ADDRESS | CITY | ZIP CODE | RES. PHONE ( ) |
| BUSINESS ADDRESS | CITY | ZIP CODE | BUS. PHONE ( ) |

| REPORTING OFFICER | ID# | DATE | REVIEWED BY | ID# | DATE |

COPIES TO: ☐ CHIEF ☐ CII ☐ PATROL ☐ DB ☐ OTHER AGENCY ROUTED BY ENTERED BY
☐ DMV ☐ CAU ☐ ABC (2 copies) ☐ DA _____

ESPD Form #269 (Rev 5/97)

El Segundo Police Department Additional Victims/Witnesses

EL SEGUNDO POLICE DEPARTMENT — ADDITIONAL VICTIMS/WITNESSES

PAGE_____ OF _____

CASE NO

CRIME 1

CODE SECTION	CRIME	CLASSIFICATION	REFER OTHER REPORTS

LOCATION	RD.	DATE	TIME	SUPPL. ☐	INCIDENT NO.

2 — WITNESS/RP/OR ADD. VICTIM

CODE	NAME (Last, First, Middle)	OCCUPATION	D.O.B.	AGE	SEX ☐ 1. M ☐ 2. F	RACE ☐ 1. WHT ☐ 2. HISP ☐ 3. BLK ☐ 5. CHI ☐ 7. FIL ☐ 9. P.ISL. ☐ 4. IND ☐ 6. JAP ☐ 8. OTH._____
RESIDENCE ADDRESS		CITY	ZIP CODE		RES. PHONE ( )	
BUSINESS ADDRESS		CITY	ZIP CODE		BUS. PHONE ( )	

(The above block of CODE / NAME / OCCUPATION / D.O.B. / AGE / SEX / RACE with RESIDENCE ADDRESS, BUSINESS ADDRESS, CITY, ZIP CODE, RES. PHONE, BUS. PHONE repeats eight times down the page.)

REPORTING OFFICER	ID#	DATE	REVIEWED BY	ID#	DATE

COPIES: ☐ CHIEF ☐ CII ☐ PATROL ☐ DB ☐ OTHER AGENCY   ROUTED BY   ENTERED BY
TO: ☐ DMV ☐ CAU ☐ ABC (2 copies) ☐ DA

ESPD Form #269 (Rev 5/97)

El Segundo Police Department Additional Victims/Witnesses

STATE OF CALIFORNIA - COUNTY OF ORANGE   SW NO. _____

SEARCH WARRANT AND AFFIDAVIT
(AFFIDAVIT)

_____ declares under penalty of perjury that the facts expressed by him/her in
(Name of Affiant)
the attached and incorporated **Statement of Probable Cause** are true and that based thereon he/she has probable cause
to believe and does believe that the articles, property,and persons described below are lawfully seizable pursuant to Penal
Code Section 1524, as indicated below, and are now located at the locations set forth below. Wherefore, affiant requests
that this Search Warrant be issued.

_____ ,   NIGHT SEARCH REQUESTED: YES [   ]   NO [   ]
(Signature of Affiant)

(SEARCH WARRANT)

THE PEOPLE OF THE STATE OF CALIFORNIA TO ANY SHERIFF, POLICEMAN OR PEACE OFFICER IN THE COUNTY OF ORANGE:
 proof by affidavit, under penalty of perjury, having been made before me by_____
—
(Name of Affiant)
that there is probable cause to believe that the property or person described herein may be found at the locations set forth
herein and that it is lawfully seizable pursuant to Penal Code Section 1524 as indicated below by "x"(s) in that it:
____ was stolen or embezzled.
____ was used as the means of committing a felony.
____ is possessed by a person with the intent to use it as means of committing a public offense or is possessed by
        another to whom he or she may have delivered it for the purpose of concealing it or preventing its discovery.
____ tends to show that a felony has been committed or that a particular person has committed a felony.
____ tends to show that sexual exploitation of a child, in violation of Penal Code Section 311.3, or possession of matter
        depicting sexual conduct of a person under the age of 18 years, in violation of Section 311.11, has occurred or is
        occurring.
____ there is a warrant to arrest the person.

YOU ARE THEREFORE COMMANDED TO SEARCH: (Premises, vehicles, persons)

FOR THE FOLLOWING PROPERTY OR PERSONS:

AND TO SEIZE IT/ THEM IF FOUND and bring it/ them forthwith before me, or this court, at the courthouse of this court. This **Search
Warrant** and Affidavit and attached and incorporated **Statement of Probable Cause** were sworn to as true under penalty of perjury
and subscribed before me on (date) _____ , at _____ A.M. / P.M. Wherefore, I find probable
cause for the issuance of this Search Warrant and do issue it.
                                                          KNOCK-NOTICE EXCUSED:□

Search Warrant and Affidavit

STATE OF CALIFORNIA - COUNTY OF ORANGE   SW NO. _____

SEARCH WARRANT AND AFFIDAVIT
(AFFIDAVIT)

_____ declares under penalty of perjury that the facts expressed by him/her in

(Name of Affiant)

the attached and incorporated **Statement of Probable Cause** are true and that based thereon he/she has probable cause to believe and does believe that the articles, property, and persons described below are lawfully seizable pursuant to Penal Code Section 1524, as indicated below, and are now located at the locations set forth below. Wherefore, affiant requests that this Search Warrant be issued.

_____, **NIGHT SEARCH REQUESTED:**   YES [ ]   NO [ ]

(Signature of Affiant)

(SEARCH WARRANT)

THE PEOPLE OF THE STATE OF CALIFORNIA TO ANY SHERIFF, POLICEMAN OR PEACE OFFICER IN THE COUNTY OF ORANGE:

proof by affidavit, under penalty of perjury, having been made before me by_____

—

(Name of Affiant)

that there is probable cause to believe that the property or person described herein may be found at the locations set forth herein and that it is lawfully seizable pursuant to Penal Code Section 1524 as indicated below by "x"(s) in that it:

_____ was stolen or embezzled.

_____ was used as the means of committing a felony.

_____ is possessed by a person with the intent to use it as means of committing a public offense or is possessed by another to whom he or she may have delivered it for the purpose of concealing it or preventing its discovery.

_____ tends to show that a felony has been committed or that a particular person has committed a felony.

_____ tends to show that sexual exploitation of a child, in violation of Penal Code Section 311.3, or possession of matter depicting sexual conduct of a person under the age of 18 years, in violation of Section 311.11, has occurred or is occurring.

_____ there is a warrant to arrest the person.

YOU ARE THEREFORE COMMANDED TO SEARCH: (Premises, vehicles, persons)

FOR THE FOLLOWING PROPERTY OR PERSONS:

AND TO SEIZE IT/ THEM IF FOUND and bring it/ them forthwith before me, or this court, at the courthouse of this court. This **Search Warrant** and **Affidavit** and attached and incorporated **Statement of Probable Cause** were sworn to as true under penalty of perjury and subscribed before me on (date) _____ , at _____ A.M. / P.M. Wherefore, I find probable cause for the issuance of this Search Warrant and do issue it.

KNOCK-NOTICE EXCUSED:☐

Search Warrant and Affidavit

_____ ,
(Signature of Magistrate)

Judge of the **Superior / Municipal** Court, _____ Judicial District, Dept. / Div. ____

NIGHT SEARCH APPROVED:☐

F026-   (Rev.1/98) Pg.1

f:\forms\sw-af1.198

STATE OF CALIFORNIA - COUNTY OF ORANGE   SW NO. _____
ATTACHED AND INCORPORATED

STATEMENT OF PROBABLE CAUSE

Affiant declares under penalty of perjury that the following facts are true and that there is probable cause to believe, and affiant does believe, that the designated articles, property, and persons are now in the described locations, including all rooms, buildings, and structures used in connection with the premises and buildings adjoining them, the vehicles and the persons:

Statement of Probable Cause

_____ ,
(Signature of Magistrate)

NIGHT SEARCH APPROVED:☐

Judge of the **Superior / Municipal** Court, _____ Judicial District, Dept. / Div. ____
__

F026-   (Rev.1/98) Pg.1

f:\forms\sw-af1.198

STATE OF CALIFORNIA - COUNTY OF ORANGE
ATTACHED AND INCORPORATED

SW NO. _____

STATEMENT OF PROBABLE CAUSE

Affiant declares under penalty of perjury that the following facts are true and that there is probable cause to believe, and affiant does believe, that the designated articles, property, and persons are now in the described locations, including all rooms, buildings, and structures used in connection with the premises and buildings adjoining them, the vehicles and the persons:

Statement of Probable Cause

STATE OF CALIFORNIA - COUNTY OF ORANGE
RETURN TO SEARCH WARRANT

_____, being sworn, says that he/she conducted a search pursuant to
(Name of Affiant)
the below described search warrant:

Issuing Magistrate: _____,

Magistrate's Court: **Superior/Municipal** Court, _____, Judicial District

Date of Issuance: _____,

Date of Service: _____,

and searched the following location(s), vehicle(s), and person(s):

R-1

Return to Search Warrant

Items attached and incorporated by Reference: YES [   ] NO [   ]
I certify (declare) under penalty of perjury that the foregoing is true and correct.

Executed at _____ , California   _____

_____   Signature of Affiant

Date:_____ at _____ [A.M.] [P.M.]

Reviewed by : _____ Date: _____ at _____ [A.M.]
[P.M.]
                    (Signature of Deputy District Attorney)

F026-   (Rev.7/97) Pg.3   Page ___ of ___

(Continued)

STATE OF CALIFORNIA - COUNTY OF ORANGE
RETURN TO SEARCH WARRANT

_____, being sworn, says that he/she conducted a search pursuant to
(Name of Affiant)
the below described search warrant:

Issuing Magistrate: _____,

Magistrate's Court: **Superior/Municipal** Court, _____, Judicial District

Date of Issuance: _____,

Date of Service: _____,

and searched the following location(s), vehicle(s), and person(s):

R-1

Return to Search Warrant

Items attached and incorporated by Reference: YES [   ]   NO   [   ]

I certify (declare) under penalty of perjury that the foregoing is true and correct.

Executed at _____ , California   _____

_____

Signature of Affiant

Date:_____ at _____ [A.M.]   [P.M.]

Reviewed by : _____ Date: _____ at _____ [A.M.]
[P.M.]

(Signature of Deputy District Attorney)

F026-   (Rev.7/97) Pg.3   Page ____ of ____

(Continued)

Glossary

Abstract Words: A form of word choice that has several meanings as opposed to a concrete word which has a clear and distinct meaning.

Abbreviations: It is generally a good idea to avoid abbreviations in report writing. Spell out the entire word because different abbreviations can mean different things to different people.

Accurate Notes: This means that the statements you write are right, measurements are precise, names are spelled correctly, and phone numbers and addresses are without error.

Active Voice: Using the active voice means you show who is doing the action before writing what they are doing.

Administrative Investigation: A noncriminal investigation typically focusing on the proprietary rights of an employer or other work rule issues.

Affiant: The person who prepares the search warrant.

Aid to Memory: Investigative notes help investigators remember what happened and assist them in their investigations. As such, notes are an aid to memory.

Automated Reporting Systems: A system that uses technology to populate record management systems with data including names, classifications, and other details of an investigation.

Average Investigator: An investigator who just gets by. Typically, they do not see the big picture and rarely try any new techniques. Usually closed to suggestions about trying something new.

Average Person Test: When describing property or evidence the description is complete enough that an average person would be able to see or hear the description and pick out the item being described.

Behavior Baseline: A general manner of behavior a person shows when truthfully answering questions during the beginning part of an interview.

Body of Data: Crime and statistical data contained in investigative reports.

Bubble In System: A form of fill in the blanks reporting. The report writer merely fills in a preformed bubble on a face sheet or other report form and a machine electronically reads the information.

Building Blocks: Investigative notes are the foundation of police reports and, as such, represent the building blocks of police reports.

Centralized Approval: A specially trained group of officers who are considered experts in report writing review and approve all reports.

Chain of Custody: Describes the handling and care of evidence from the time it is discovered until it is introduced in court.

Check the Box: A form of forced choice used in completing crime and arrest report face sheets.

Chronological Order: Describes the process of writing about what happened in the order it happened.

Computer: From the report writing perspective, computers have had an enormous impact. The use of computers allows many cross applications of technology to the needs of law enforcement.

Concise: Using as few words as possible to accurately record what happened.

Concrete Words: Using the lowest level of abstraction in describing something. Concrete words are the opposite of Abstract words.

Corpus Delicti: The elements or body of the crime.

Credibility: This is what investigators gain when they do a thorough and professional job, it is tantamount to having a good reputation.

Crime Report: A general term for the report that contains the elements of a crime and establishes that a crime occurred.

Criminal Investigation: Typically conducted by law enforcement personnel and may not begin until a crime has occurred, there is a reasonable certainty a crime has occurred or the investigator is reasonably sure a crime is going to occur.

Dictation: When a report preparer reads or says something that another person later transcribes or writes down.

Direct Quotes: Writing exactly what a person said.

Drawing: A technique in which a victim or witness actually draws a picture of something. This can be very valuable information when trying to match stolen property against a description or list.

Evidence List: A description of the evidence seized during an investigation. The evidence may be listed in the order it was seized or by using general categories such as Guns, Money, Drugs, Items with serial numbers and Items without serial numbers.

Evidence Report: A report that describes the evidence in detail and describes the chain of Custody.

Exhibits: When attachments are included in a search warrant they are referred to as exhibits.

Expert Opinion: This is the opinion of someone who has extensive experience, training and education about a particular area. The combination of this depth and breadth about a subject qualifies the person to offer an expert opinion about something. Generally, for report writing purposes, an expert opinion should be included in a report only when it establishes probable cause.

Face Sheet: A part of a crime report that is mainly composed of Fill in the blanks or Forced Choice cells.

Fact: Something that can be proven. Police reports should be composed of facts, not opinions.

Fair Market Value: When investigators use their judgment in determining the value of an item by considering what it was worth when it was new, what the demand for the item is, and what it might be worth at the time of the report. This is known as the Fair Market Value.

Field Identification: A process where a victim or witness is allowed to see suspects in the field for the purpose of identifying the person as the perpetrator of the crime, or to eliminate the person from the suspect pool. This process is also referred to as a Show Up.

Fill in the Blanks: Generic information like name, address, and phone number. This is typically information that needs to be included on a report face sheet.

First Person: The recommended method of writing an investigative report. When writing in the First Person the writer refers to himself as "I".

Fixed Terminal: This is a computer that is hard wired into a car or a work station.

Forced Choice: Preselected choices that a writer uses to select the appropriate condition or descriptor in a fill in the blank portion of a report.

Gathering Statistics: One of the two major purposes of a crime report face sheet. The other is to organize information.

Grammar Check: A feature of writing software that allows writers to check the proposed language of a narrative against standard usage and offers suggestions when a questionable situation occurs.

Indirect Quotes: Refers to the process of paraphrasing what someone says.

Interview: A facilitative process in which the investigator asks questions and receives information from the person being questioned.

Interrogation: A focused question and answer session in which there is a higher level of hostility and anxiety than in an interview.

Interview Worksheet: A tool investigators use to prepare for an interview. The worksheet prompts the investigator to find out as much background information as possible about the person to be interviewed before the interview begins.

Investigation: The work done to find out what happened. A definition of an investigation is that it is a lawful search for things or people.

Investigative Tool: The most important use of a police report.

Investigator: The person who looks into events or situations to find the facts about what happened.

Invoking Rights: After a suspect is told his Miranda rights and declines to talk to the interviewer, he is said to have invoked his rights.

Jargon: Slang words or phrases that may be specific to a particular law enforcement agency. They should not be used in police reports.

Key Words or Phrases: Generally refers to the words a suspect uses during the commission of a crime and what a suspect says when confessing.

Listen First, Then Write: When taking notes it is important to let the person talk before you start writing things in your notebook. This refers to the technique of letting the person tell you the story before you put it into writing.

Measurements: Accurate measurements are critical to the note taking and investigative process.

Mechanics of Note Taking: The practical application of writing down the initial findings of an investigation in a notebook.

Miranda Admonishment: This refers to the rights a person has against self incrimination. Preferably, the Miranda Admonishment is read to a suspect and the suspect's response is quoted.

Names and Titles: Refer to people in a report by their name and avoid using a title like victim, suspect, or witness.

Narrative: The free flowing part of the report where the writer tells the story of what happened, what actions were taken to solve the crime and any evidence that was collected.

Narrative Section: The part of a report in which the writer tells the story of what happened in a free flowing, continuous written format.

National Incident Based Crime Reporting System: NIBRS was created in a 1982 overhaul of the Uniform Crime Reporting Program and in addition to capturing the Summary Data it always did, it also captures more specific details about each crime such as date, time, location, circumstances of the incident, the sex, race and age of the victim and suspect, any relationship information between them, whether drugs or weapons were involved and any evidence of whether the crime was motivated by bias.

New York State Incident Based Reporting Program: NYSIBR, modeled after the federal NIBRS program, this model goes further in standardizing reporting. This program includes standardized incident and arrest reports, views an incident as a set of related components, looks at relationships between suspects, victims and property, victim and suspect demographics, use of weapons and whether drugs and alcohol were involved.

Non-Verbal Responses: The actions, facial expressions and body movements a person makes while answering questions in an interview.

Opinion: A personal belief or judgment that cannot be proven.

Optical Character Recognition: OCR translates scanned images of handwritten or typewritten words into computer edible text.

Organize Information: One of the two main purposes of a crime report face sheet. The other one is to gather statistics.

Original Cost Method: A way of determining value for a piece of property. The cost of the item when new is used to determine this value.

Others: People who have no information about a case, but were contacted by the investigator. They may have been in the area of the crime and were talked to by the police.

Part I Crimes: Murder, rape, robbery, aggravated assault, burglary, larceny-theft, motor vehicle theft, and arson.

Past Tense: The verb tense investigative reports should be written in.

Probable Cause: Refers to the amount of proof or belief that must exist before an arrest is made. The probable cause is based on the facts and evidence known at the time an arrest is made.

Pull Down Menu:　Much like a forced choice option. With pull down menus, the writer clicks on an icon and a preprogrammed list of choices appears. The writer then selects the one needed.

Purpose of a Crime Report Face Sheet:　The two purposes of a crime report face sheet are to organize information and gather statistics.

Radio Code:　While radio code is used in general conversation, it should not be used in police reports. This would be considered jargon if included in the narrative of the report.

Readable:　One of the few rules that apply to note taking. Your notes need to be legible and readable.

Reader Use Conflict:　People who are authorized to read and use information in police reports may have different reasons for doing so. Not everyone sees the facts and details of an investigation in the same light. When the reasons that people have for using a police report differ, they have a conflict in what they want the report to do or be. This is known as reader use conflict.

Reasonable Particularity:　When writing a search warrant this is the test of how good a location description must be. A description must be so complete that any police officer could read the description of the premises to be searched and find it.

Records Management System:　A database of all investigative data generated including, demographic information, names of victims, witnesses and suspects as well as dates and times of incidents. The records management system allows searches to be run by investigators looking for trends, patterns and historical data.

Replacement Cost Method:　This is what it will cost to replace an item in kind.

Report Formats:　When a specific layout or order of information is not specified by an agency, the writer must develop a style or format of organizing and documenting the facts of the case. This is the report format and may vary from one agency to another.

Report Writing Issues:　A term that describes potential problems in the investigative and writing phases of a case.

Reporting District:　A geographical area used to track calls for service, reported crimes and arrests. The reporting district is typically the basis for all police statistics.

Reporting Party:　The person who reports the crime to an investigative agency. They are sometimes referred to as the Person Reporting.

Return to the Warrant:　The part of the search warrant that contains the list of evidence seized during the search.

Rules of Narrative Writing:　Refers to a writing system that includes writing in the first person, using the past tense, writing with the active voice, listing events in chronological order starting with the date, time and how you got involved and using short, clear, concise and concrete words.

Search Warrant:　A search warrant consists of three parts, the search warrant, the affidavit in support of the warrant and the return to the warrant. In general it is an order from a judge to a peace officer to go to a particular place, look for a particular thing and if it is found, bring it back to the court. The search warrant is also the part that lists the address to be searched and a description of the evidence you are looking for.

Search Warrant Affidavit:　Also known as the Statement of Probable Cause, it contains the information and facts that establish the probable cause to justify the issuance of the warrant.

Show Up:　A process where a victim or witness is allowed to see suspects in the field for the purpose of identifying the person as the perpetrator of the crime, or to eliminate the person from the suspect pool. This process is also referred to as a Field Identification.

Sketch:　A crime scene sketch can be a valuable tool in an investigation. If video or still picture resources are not available, a simple drawing with measurements and important details will be very useful.

Solvability Factors:　General questions about a case that can be answered with a yes or no. These Factors, as they are sometimes called, are used to prioritize cases and identify the ones with the greatest chance of being solved.

Spell Check: Technology that allows writers to automatically check the spelling of words in their reports against a standard list.

Spelling: A common problem for report writers is that words are misspelled. The words a writer chooses must be spelled correctly.

Statement of Probable Cause: See Search Warrant Affidavit.

Storage: An investigator's notes represent the first level of records storage.

Summary Data Report System: On a monthly basis, 17,000 law enforcement agencies report statistical data to the FBI on a voluntary basis. This data is used to create the Uniform Crime Reports.

Superior Investigator: These are the investigators who are willing to go the extra mile, try new things, and look at a set of facts with an open and inquisitive mind. They get results by doing all the little things, and doing them the right way.

Supplemental Reports: A catch all term for the reports that contain all the information that does not fit in a specific report.

Suspect: When the evidence or information justifies the arrest of a particular person, that person is said to be the suspect. Generally speaking, the suspect is believed to be the person responsible for committing the crime.

Template: A form used in report writing that has preprogrammed fill in the blank and forced choice options. A template also standardizes the information captured in a report.

Thirty Word Version: An abbreviated version of what happened. Typically a witness is asked to describe what happened in "thirty words."

Touch Screen Features: Allow for the rapid selection of common data. Usually they are set up to select the type of report, date and time, distribution and recommended follow-up.

Transcribe: When someone makes a written copy of dictated material.

Uniform Crime Reporting Program: A program managed by the FBI that collects and tracks statistics on the eight Part I Crimes.

Victim: Someone who has been hurt or who may have had their property damaged or stolen.

Victim Appraisal Method: The victim provides the value of the stolen or damaged property.

Verbal Response: The things a person says in response to interview questions.

Voice Dictation: Technology that uses a preprogrammed voice recognition feature. The report composer dictates the report and the software, after identifying and recognizing the voice converts it to transcribed text.

Waiver: The process in which a suspect receives his Miranda rights and agrees to talk to the interviewer.

Witnesses: Someone who has useful information about the crime you are investigating.

Index